DRILLING CALIFORNIA

A Reality Check on the Monterey Shale

By J. David Hughes
December 2013

post carbon institute

PSE
Physicians Scientists & Engineers for Healthy Energy

About the Author

J. David Hughes is a geoscientist who has studied the energy resources of Canada for nearly four decades, including 32 years with the Geological Survey of Canada as a scientist and research manager. He developed the National Coal Inventory to determine the availability and environmental constraints associated with Canada's coal resources. As Team Leader for Unconventional Gas on the Canadian Gas Potential Committee, he coordinated the recent publication of a comprehensive assessment of Canada's unconventional natural gas potential. Over the past decade, Mr. Hughes has researched, published, and lectured widely on global energy and sustainability issues in North America and internationally.

In 2011, Mr. Hughes authored a series of papers on the production potential and environmental impacts of U.S. natural gas. In early 2013, he released *Drill, Baby, Drill: Can Unconventional Fuels Usher in a New Era of Energy Abundance?*, which takes a far-ranging and painstakingly researched look at the prospects for various unconventional fuels to provide energy abundance for the United States in the 21st century. While the report examines a range of energy sources, the centerpiece of *Drill, Baby, Drill* is a critical analysis of shale gas and tight oil (shale oil) and the potential of a shale "revolution."

Mr. Hughes is president of Global Sustainability Research, a consultancy dedicated to research on energy and sustainability issues. He is also a board member of Physicians, Scientists & Engineers for Healthy Energy (PSE Healthy Energy) and the Association for the Study of Peak Oil and Gas – Canada, and is a Fellow of the Post Carbon Institute. Mr. Hughes recently contributed to *Carbon Shift*, an anthology edited by Thomas Homer-Dixon on the twin issues of peak energy and climate change, and his work has been featured in *Nature*, *Canadian Business*, and other journals, as well as through the popular press, radio, and television.

About Post Carbon Institute

Post Carbon Institute builds resilience by providing individuals, communities, businesses, and governments with the resources needed to understand and respond to the interrelated economic, energy, environmental, and equity crises that define the 21st century. **postcarbon.org**

About Physicians Scientists & Engineers for Healthy Energy

As a team of leading scientists, engineers, and health professionals, PSE Healthy Energy brings scientific transparency to important energy policy issues and energy choices. PSE Healthy Energy generates, translates, and disseminates scientific information on the climate, public health, and environmental dimensions of oil and gas production, the transition to renewable energy portfolios, and other novel forms of energy production. **psehealthyenergy.org**

Acknowledgments

Thanks to Asher Miller and Daniel Lerch of Post Carbon Institute and Seth Shonkoff of Physicians Scientists & Engineers for Healthy Energy for editing and reviewing the report, and to John Van Hoesen of Green Mountain College for designing and producing the maps.

Drilling California: A Reality Check on the Monterey Shale
By J. David Hughes
In association with Post Carbon Institute and Physicians Scientists & Engineers for Healthy Energy
Copyright © 2013 by J. David Hughes. All rights reserved. Published December 2013.

For reprint requests and other inquiries, please contact Post Carbon Institute,
613 Fourth St., Suite 208, Santa Rosa, California, 95404

EXECUTIVE SUMMARY

The recent growth in unconventional oil production from the Bakken (North Dakota), Eagle Ford (Texas), and other tight oil plays has drawn attention to the potential of shale in California's Monterey Formation. Commercial oil production from the Monterey Formation is not new—more than a billion barrels of oil and four trillion cubic feet of gas have been produced from it since 1977, largely from conventional reservoirs. However, completion techniques like hydraulic fracturing which have made tight oil production possible from shale deposits elsewhere have not yet been widely implemented in the shale source rocks of the Monterey Formation.

In 2011, the U.S. Energy Information Administration (EIA) published a report by INTEK Inc. which stated that the Monterey Formation contains 15.4 billion barrels of technically recoverable tight oil (therein referred to as "shale oil")[1]—64 percent of the entire estimated tight oil resource in the Lower-48 United States at that time. This estimate was seized upon by industry groups intent on the development of the Monterey shale, and was used as the basis of a March 2013 University of Southern California (USC) economic analysis which projected as much as a $24.6 billion per year increase in tax revenue and 2.8 million additional jobs by 2020.[2] It also raised alarm among groups concerned about the environmental and public health implications of hydraulic fracturing, acidization, and other advanced well stimulation technologies.

This report provides the first publicly available empirical analysis of oil production data from the Monterey Formation, utilizing the Drillinginfo database (widely used by the oil and gas industry as well as the EIA). It lays out some of the fundamental characteristics of the Monterey compared to other tight oil plays, including geological properties, current production, and production potential. The results of this analysis will be useful for informing public policy decisions surrounding the development of the Monterey shale.

Prospects for Tight Oil Production in the Monterey Formation

Much of the enthusiasm about developing tight oil in the Monterey comes from the success of tight oil production in the Bakken and the Eagle Ford plays of North Dakota/Montana and southern Texas, respectively. Geologically, however, these two plays are very different from the Monterey:

- The target strata in the Bakken and the Eagle Ford plays are less than a few hundred feet in thickness and are flat-lying to gently dipping. The shale deposits of the Monterey are much thicker and much more complex, with target strata up to 2,000 or more feet in thickness, and at depths that can range from surface outcrops to more than 18,000 feet within a span of forty miles or less.

- The Bakken play is spread over a potentially productive area of as much as 20,000 square miles, and the Eagle Ford play covers roughly 8,000 square miles. The Monterey tight oil play encompasses less than 2,000 square miles.

---

[1] INTEK, Inc., *Review of Emerging Resources: U.S. Shale Gas and Shale Oil Plays*, December 2010, in U.S. Energy Information Administration, *Review of Emerging Resources: U.S. Shale Gas and Shale Oil Plays*, July 2011, http://www.eia.gov/analysis/studies/usshalegas/.
[2] University of Southern California, USC Price School of Public Policy, *The Monterey Shale and California's Economic Future*, (March 2013), http://gen.usc.edu/assets/001/84955.pdf.

- The Bakken and Eagle Ford shales are approximately 360 and 90 million years old, respectively, and were deposited on a relatively stable platform; as a result they are thin and widespread, and hence are relatively predictable. The Monterey is 6-16 million years old and was deposited rapidly in an active tectonic regime; as a result it is thick, of limited areal extent, and structurally complex, and hence is much less predictable.

An analysis of oil production data from the Monterey Formation reveals the following:

- While tight oil plays generally produce directly from widely dispersed source rocks or immediately adjacent reservoirs—as is the case in the Bakken and the Eagle Ford—this is not the case in the Monterey Formation, where most production has come from localized conventional reservoirs filled with oil that has migrated from source rock.

- 1,363 wells have been drilled in shale reservoirs of the Monterey Formation. Oil production from these wells peaked in 2002, and as of February 2013 only 557 wells were still in production.[3] Most of these wells appear to be recovering migrated oil, not "tight oil" from or near source rock as is the case in the Bakken and Eagle Ford plays.

The EIA/INTEK report assumed that 28,032 tight oil wells could be drilled over 1,752 square miles (16 wells per square mile) and that each well would recover 550,000 barrels of oil. The data suggest, however, that these assumptions are extremely optimistic for the following reasons:

- Initial productivity per well from existing Monterey wells is on average only a half to a quarter of the assumptions in the EIA/INTEK report. Cumulative recovery of oil per well from existing Monterey wells is likely to average a third or less of that assumed by the EIA/INTEK report.

- Existing Monterey shale fields are restricted to relatively small geographic areas. The widespread regions of mature Monterey shale source rock amenable to high tight oil production from dense drilling assumed by the EIA/INTEK report (16 wells per square mile) likely do not exist.

Thus the EIA/INTEK estimate of 15.4 billion barrels of recoverable oil from the Monterey shale is likely to be highly overstated. Certainly some additional oil will be recovered from the Monterey shale, but this is likely to be only modest incremental production—even using modern production techniques such as high volume hydraulic fracturing and acidization. This may help to temporarily offset California's long-standing oil production decline, but it is not likely to create a statewide economic boom.

Why These Findings are Important

The March 2013 study from the University of Southern California[4] suggested that the development of the Monterey shale could—by 2020—increase California's Gross Domestic Product (GDP) by 14 percent, provide an additional 2.8 million jobs (a 10% increase), and provide $24.6 billion per year in additional tax revenue (also a 10% increase). This study was based on estimates that development of the Monterey shale could increase total California oil production as much as seven-fold. Given the unrealistic nature of the original EIA/INTEK Monterey shale estimates, such production growth estimates are unfounded. Moreover, an examination of USC's oil production estimates reveals that they

---

[3] Data from DI Desktop (Drillinginfo).
[4] University of Southern California, USC Price School of Public Policy, *The Monterey Shale and California's Economic Future*, (March 2013), http://gen.usc.edu/assets/001/84955.pdf.

include unrealistic assumptions about the total production growth possible from the Monterey and the number of wells that would be required to increase production to the levels forecast. Hence the economic projections of the USC study must be viewed as extremely suspect.

Environmental concerns with development of the Monterey shale are centered around hydraulic fracturing (fracking), the main completion technique used in other tight oil plays. Acidization completions, using hydrofluoric and hydrochloric acid, are also of concern. It is certain that hydraulic fracturing and acidization completions have already been used on the Monterey shale, yet an analysis of production data reveals little discernible effect of these techniques in terms of increased well productivity. Many oil and gas operators and energy analysts suggest that it is only a matter of time before "the code is cracked" and the Monterey produces at rates comparable to the Bakken and Eagle Ford. Owing to the fundamental geological differences between the Monterey and other tight oil plays, and in light of actual Monterey oil production data, this is likely wishful thinking.

For all of these reasons, this analysis suggests that California should consider its economic and energy future in the absence of an oil production boom from the Monterey shale.

CONTENTS

FIGURES

TABLES

1 INTRODUCTION

Development of tight oil (also known as "shale oil") from shale reservoirs in North Dakota (Bakken), southern Texas (Eagle Ford), and elsewhere in the past few years has allowed the U.S. to temporarily reverse its long-standing oil production decline that began in 1970. This has been made possible by technological advances allowing high-volume, multi-stage hydraulic fracturing of horizontal wells, as well as other innovations. The success of these practices has prompted calls that the U.S. is becoming "Saudi America," and predictions that the country will soon become "energy independent" and even a net oil exporter (despite the fact that the U.S. currently is the second largest oil importer in the world, after China).

The Monterey Formation[5] of south-central California is considered a major potential source of tight oil (Figure 1). It has been touted as an economic bonanza, but also as a potential environmental disaster given that production techniques would involve both hydraulic fracturing (fracking) and acidization. The prize is said to be as much as 15.4 billion barrels of technically recoverable oil—64 percent of the entire estimated tight oil resource in the Lower-48 United States as of 2011.[6] Concerns surrounding recovery of this oil relate to the many real and perceived environmental issues associated with tight oil (and shale gas) recovery elsewhere in the U.S.

In fact, the Monterey Formation has produced oil (largely from conventional reservoirs) for many decades. It is also an important source rock; oil migrating from it has charged many large conventional fields in the San Joaquin, Santa Maria, and Ventura Basins with significant volumes of both oil and gas. As in the Bakken and Eagle Ford, tight oil production in the Monterey Formation targets the source rocks and their associated reservoirs. The geology of the Monterey is vastly different from that of the Bakken and Eagle Ford, however, and poses significant challenges that do not appear to have been duly considered in the most optimistic production forecasts.

This report provides the first publicly available empirical analysis of actual oil production data from the Monterey Formation. It lays out some of the play's fundamental characteristics compared to other tight oil plays, including geological properties, current production, and production potential. It will be useful for informing public policy decisions on the developing the Monterey.

---

[5] Throughout this report, all references to the Monterey should be understood to include the Santos as well.
[6] INTEK, Inc., *Review of Emerging Resources: U.S. Shale Gas and Shale Oil Plays*, December 2010, in U.S. Energy Information Administration, *Review of Emerging Resources: U.S. Shale Gas and Shale Oil Plays*, July 2011, http://www.eia.gov/analysis/studies/usshalegas/.

Figure 1. The Monterey tight oil play in California, with relevant sedimentary basins and counties.

The portions of the Monterey Formation that are thought to have potential for tight oil production lie in the area delineated as the play.[7] Most production has been in relatively small portions of the sedimentary basins found within the play. The extent of the entire Monterey Formation is not shown.

---

[7] See also Figure 7. The boundaries of the play are approximate and based on the area marked as such in INTEK, Inc., *Review of Emerging Resources: U.S. Shale Gas and Shale Oil Plays*, December 2010, in U.S. Energy Information Administration, *Review of Emerging Resources: U.S. Shale Gas and Shale Oil Plays*, July 2011, http://www.eia.gov/analysis/studies/usshalegas/.

1.1 Method

This report evaluates the claim of a 2011 report released by the U.S. Energy Information Administration and prepared by INTEK Inc. (EIA/INTEK) that tight oil production in the Monterey Formation could ultimately yield 15.42 billion barrels of oil. Central to the EIA/INTEK report's assumptions are that:

a.) The potential for tight oil production in the Monterey is analogous to other tight oil plays like the Bakken and Eagle Ford.

b.) The inferred tight oil production potential can be applied uniformly to the entire play, without regard to the widely varied geological characteristics within the Monterey Formation.

Therefore, this report:

a.) Reviews the production history of the Monterey Formation, which has largely occurred in the southern San Joaquin Basin and the Santa Maria Basin (it excludes the Ventura Basin and Los Angeles Basin because there is currently no significant onshore oil production from the Monterey there). New offshore production is also excluded since this is unlikely to take place in the foreseeable future because of a five-year ban on West Coast offshore drilling imposed in 2011, as well as historically strong opposition to California offshore drilling following the 1969 Santa Barbara oil spill.

b.) Reviews the geological characteristics of the onshore Monterey Formation.

This report also assesses the claim of a 2013 study published by the University of Southern California (USC) that tight oil production in the Monterey Formation would greatly boost statewide employment, per capita GDP, and tax revenue. The Monterey oil production assumptions on which the USC study is based are examined in the light of actual production data.

The well, reservoir, and field production data analyzed in this report is sourced from the DI Desktop software package provided by Drillinginfo (formerly HPDI), which is widely used by the oil and gas industry and government as an authoritative source for oil and gas production information in North America. This database, which is updated monthly, allows oil and gas production data to be analyzed over time by well, reservoir, operator, county, well type, and other variables. The online data and GIS mapping capabilities of the California Division of Oil, Gas and Geothermal Resources (DOGGR) were also utilized. The scope of this analysis includes all wells recorded to have been drilled in the Monterey Formation from 1977 to mid-2013.

A Brief Primer on Terminology and Oil Generation

Terminology

Formation is a formal name for a rock unit that can be recognized over relatively large geographic areas. A formation can be subdivided into *members*, and included with other formations in a group. The Monterey Formation, for example, is subdivided into members such as the Stevens Sand, McLure Shale, Reef Ridge Shale, Antelope Shale, and so forth.

The Monterey Formation contains both *source rocks*, where oil has been generated, and *reservoir rocks*, where oil that has migrated from the source rocks is trapped by a seal, either diagenetic or structural (fault or fold). In some cases, such as tight oil plays like the Bakken and Eagle Ford, source rocks are also reservoir rocks, as the oil contained within them has migrated little or no distance owing to very limited permeability.

The Oil and Gas Generation Process

1. Sediments with sufficient organic content (total organic carbon) accumulate and over time are buried by *sedimentation* and *tectonic activity*.

2. At sufficient depth (approximately 2-4 km) these sediments enter the *oil generation window*, an interval in the subsurface where temperatures and pressures are high enough for organic matter to undergo thermogenic breakdown (cracking), generating oil over sufficient periods of time. At yet greater depths of burial (3-6 km) and correspondingly higher temperatures, gas is generated.

3. Oil or gas may then be expelled from the source rock and migrate through permeable rocks or fractures until it is trapped by a tight, non-permeable layer of rock (like a shale) and creates a reservoir—which may then be tapped by a well as *a conventional oil play*. If the oil or gas is not trapped, it may migrate to the surface. After burial and hydrocarbon generation, source rocks may also be uplifted by tectonic forces.

4. Unconventional *tight oil plays* tap oil remaining in the source rocks themselves, or in immediately adjacent tight rocks, by inducing permeability through the creation of artificial fractures (hydraulic fracturing) or through dissolution of the rock matrix (acidization) through which oil can migrate to the well bore.

2 THE MONTEREY IN CONTEXT

2.1 California Oil Production

Although California has produced oil and natural gas for more than a century, statewide production peaked in 1985 at 1.1 million barrels per day (mbd) and has declined by more than 50 percent since then (Figure 2). Since 1977, which is the earliest date for which digital well production data are available, some 10.4 billion barrels of oil and 17.1 trillion cubic feet of natural gas have been produced in the state.[8] To put this in perspective, the U.S. consumes approximately 6.8 billion barrels of oil each year along with 24 trillion cubic feet of gas.[9] In other words, over the past 30 years, California has produced enough oil and gas to satisfy U.S. demand for approximately 1.5 years for oil and 0.7 years for gas.

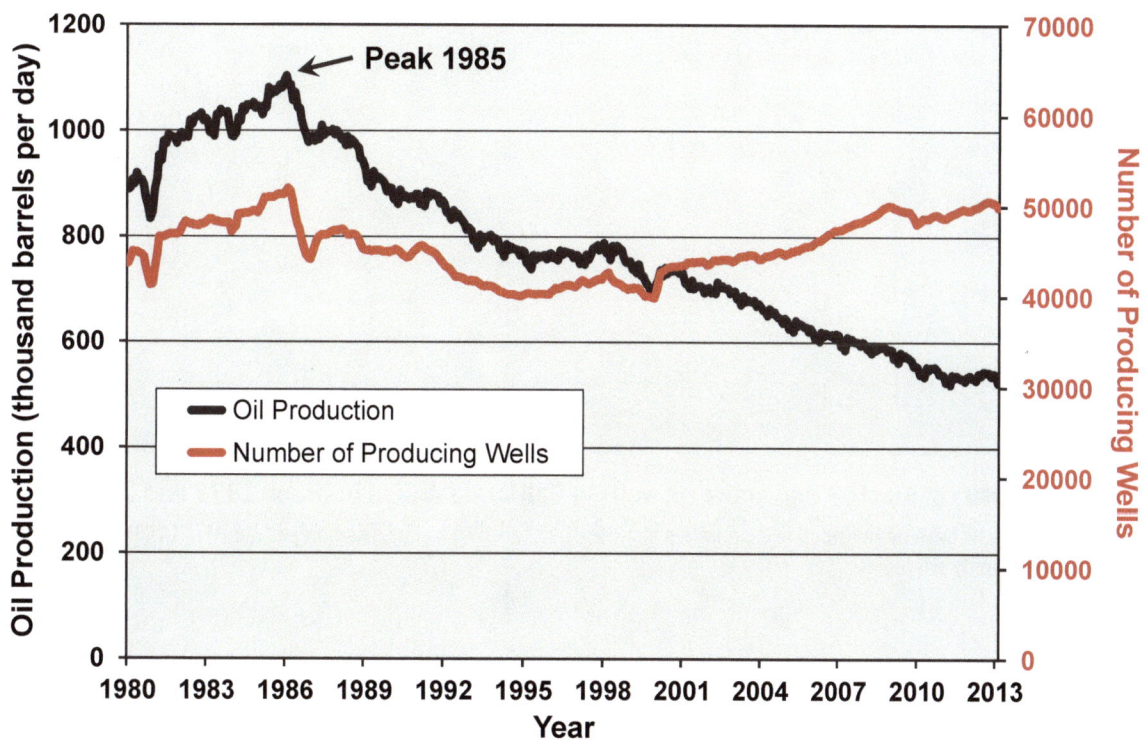

Figure 2. California oil production and number of producing wells, 1980 through May 2013.[10]

---

[8] Data from DI Desktop (Drillinginfo), current through May 2013.
[9] U.S. Energy Information Administration, 2013.
[10] Data from DI Desktop (Drillinginfo), current through May 2013.

Although overall oil production in California has decreased by more than half since 1986, the number of operating wells has remained roughly constant. More specifically, production per oil well has decreased from an average of 22 barrels per day in 1986 to just over 10 barrels per day in early 2013. Today, approximately 50,000 wells are contributing to current production, but over 238,000 wells have been drilled, as illustrated in Figure 3.

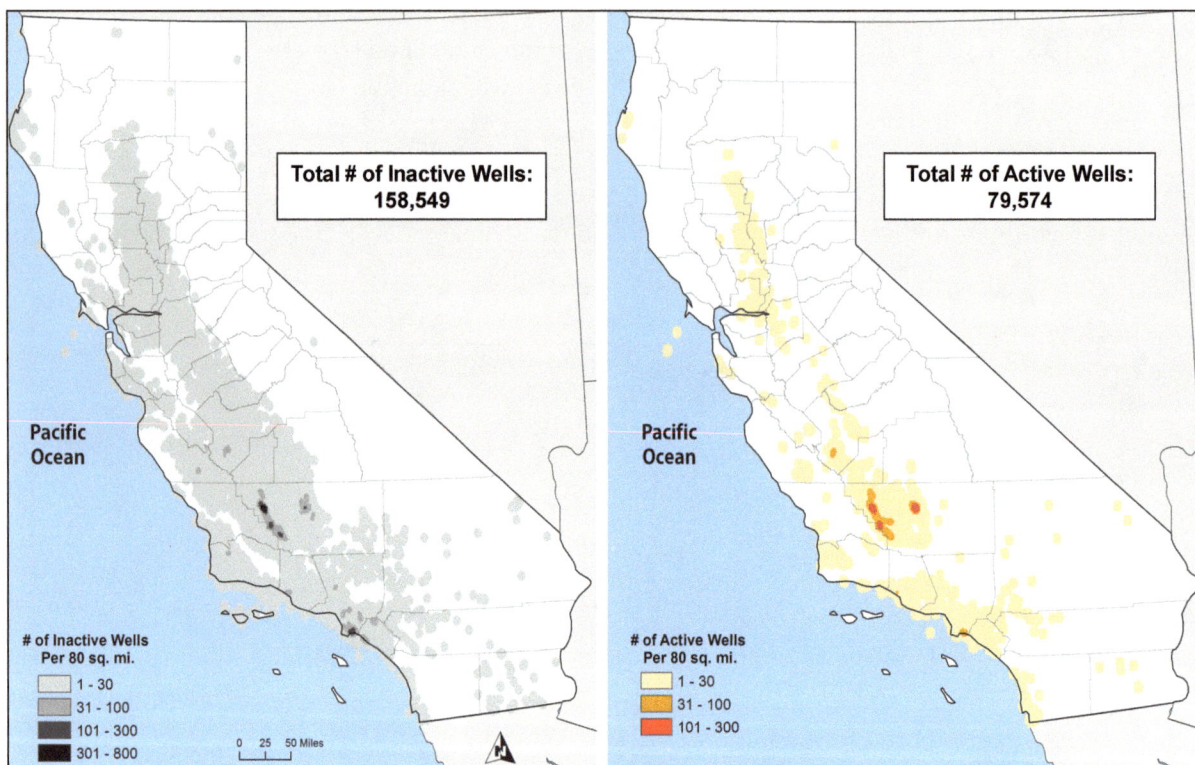

Figure 3. Density of inactive and active oil wells in California drilled between 1977 and 2013.[11]

Over 238,000 wells have been drilled in California's history. Of the 79,574 wells classified as "Active or Drilling," only about 50,000 are currently producing.

[11] Data from DI Desktop (Drillinginfo), current through May 2013.

Kern County is by far the largest recipient of this drilling activity, followed by Los Angeles, Fresno, Orange and other counties as illustrated in Figure 4.

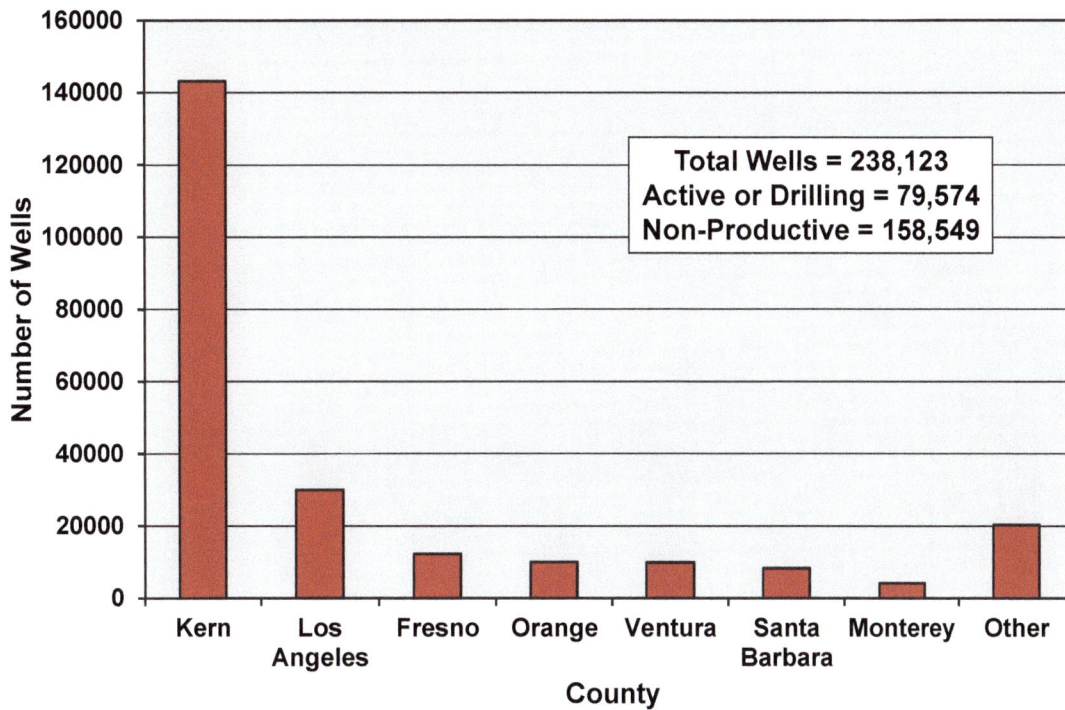

Figure 4. Number of oil and gas wells drilled in California, by county through May 2013.[12]

Of the 79,574 wells classified as "Active or Drilling," only about 50,000 are currently producing.

---

[12] Data from DI Desktop (Drillinginfo), current through May 2013.

As average well production declined, the industry shifted to unconventional methods such as cyclic steam injection to recover lower quality, heavier gravity oil. Cyclic steam injection involves the injection of steam into a well to fracture and heat the reservoir rock. This in turn reduces the viscosity of the oil to allow it to flow to the well bore and be collected at the wellhead. As illustrated in Figure 5, the share of California oil production accounted for by cyclic steam wells increased from 11 percent in 1980 to over 20 percent today—largely because production using conventional oil recovery methods has decreased.

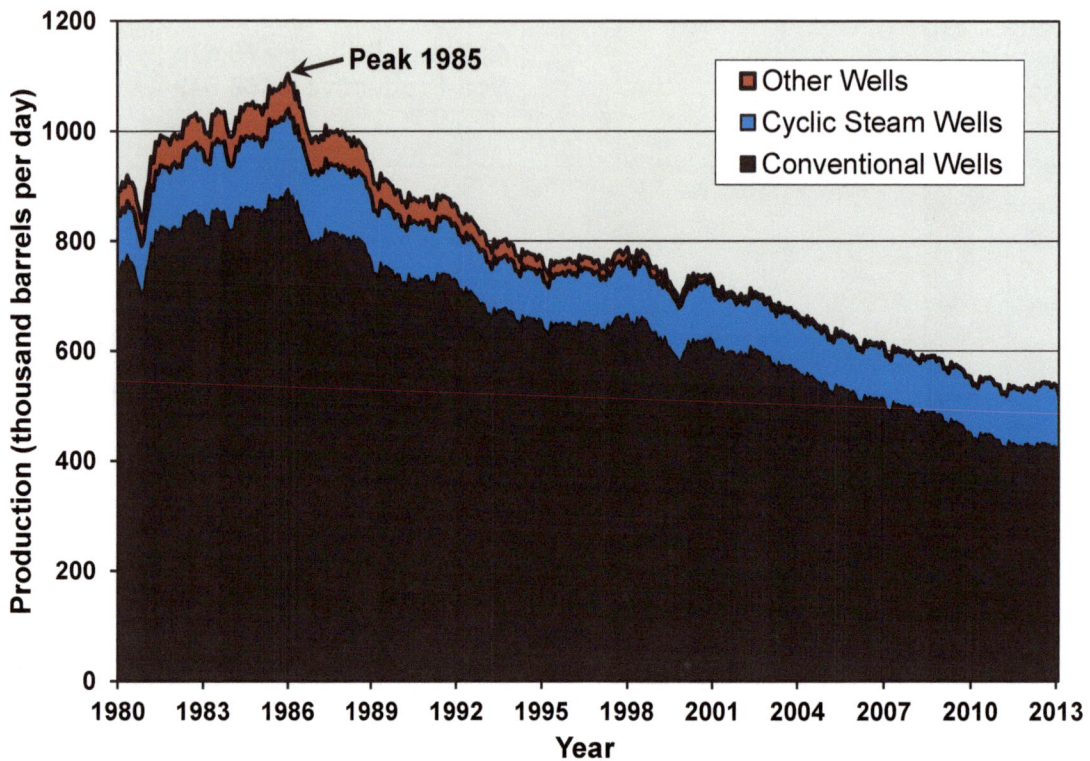

Figure 5. California oil production by well type, 1980 through May 2013.[13]

"Other Wells" include those designated as "gas" and those drilled for other purposes such as water disposal and pressure maintenance.

Thus California's oil production story has been one of inexorable decline, along with a decline in associated economic benefits generated by the oil and gas industry.

---

[13] Data from DI Desktop (Drillinginfo), current through May 2013.

The recent rapid growth of U.S. tight oil production—made possible by the advent of high-volume, multi-stage, hydraulic fracturing of horizontal wells—has sparked hope for similar developments in California. Interest has centered on the Monterey Formation in the south-central part of the state.

In fact, the Monterey Formation has been producing oil in California for decades (Figure 6), and shales within it are the source rocks that charged many of California's conventional oil reservoirs. It is a young formation (of Miocene age, approximately 6-16 million years old) that was deposited in an active tectonic setting and geologically has little in common with the tight oil plays that have been successfully developed elsewhere (see Section 4.1). Thousands of Monterey wells have been drilled in Kern and Santa Barbara counties (including offshore), generally without the stellar results that characterize the best tight oil plays, calling into question the assumptions that underlie the EIA/INTEK report estimates.

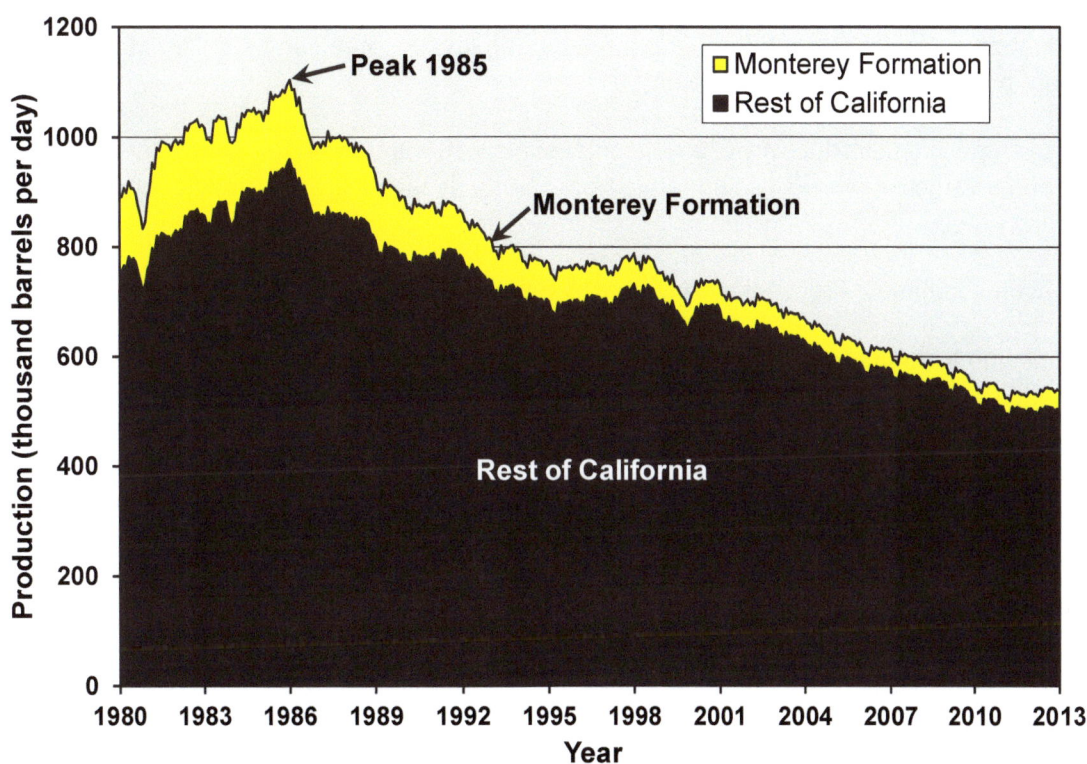

Figure 6. Oil production from the Monterey Formation and from the rest of California, 1980 through May 2013.[14]

California production peaked in 1985; Monterey Formation production peaked in 1982.

---

[14] Data from DI Desktop (Drillinginfo), current through May 2013; three-month trailing moving average.

2.2   The EIA/INTEK Report and USC Study

In 2011 the U.S. Energy Information Administration released a report by INTEK Inc. (hereafter the "EIA/INTEK report") which stated that as a tight oil play (i.e., not including conventional production) the Monterey had 15.42 billion barrels of technically recoverable oil[15]—an amount equal to 64% of the estimated total tight oil resource in the Lower-48 United States. The report created a great amount of interest in the Monterey, as well as concern about the environmental and public health impacts of its development. The EIA/INTEK estimate of 15.42 billion barrels was based on the following assumptions:

- The Monterey tight oil play encompasses 1,752 square miles, averaging 11,250 feet in depth and 1,875 feet in thickness.

- Production potential is found throughout the entire play; the area can be drilled at 16 wells per square mile, for a total of 28,032 wells.

- The average initial productivity of a horizontal shale well will be 500 barrels of oil equivalent per day and a vertical shale well could range from 250 to as high as 800 barrels of oil equivalent per day.

- The average ultimate recovery from each well will be 550,000 barrels of oil.

- Completion technologies that have been successful in other tight oil plays such as hydraulic fracturing will produce similarly high-yield results.

The EIA later lowered the report's estimates slightly to 13.7 billion barrels, and the average production per well to 497,000 barrels, but still kept the report's assumptions of 28,000 wells covering 1,750 square miles (16 wells per square mile).[16] However, virtually all public attention has remained focused on the higher estimate of 15.42 billion barrels.

Then in March 2013, the University of Southern California (USC) released a study suggesting that development of the Monterey shale could—by 2020—increase California's GDP by 14 percent, provide an additional 2.8 million jobs (a 10% increase), and provide $24.6 billion per year in additional tax revenue (also a 10% increase). The study was based on production growth forecasts that even it admitted were "possibly very optimistic".[17]

A credible assessment of California's energy and economic options going forward is essential and—in the case of tight oil production from the Monterey—is best understood in the context of the geology, production history, and the likely future production potential, along with the environmental issues that may be encountered in its development. As this report will show, an analysis of actual production data from the Monterey calls into question the assumptions that underlie the EIA/INTEK report estimates, and the credibility of projections of economic benefit that are based on them.

---

[15] INTEK, Inc., *Review of Emerging Resources: U.S. Shale Gas and Shale Oil Plays*, December 2010, in U.S. Energy Information Administration, *Review of Emerging Resources: U.S. Shale Gas and Shale Oil Plays*, July 2011, http://www.eia.gov/analysis/studies/usshalegas/. The report refers to the "Monterey/Santos."
[16] U.S. Energy Information Administration, *Assumptions to the Annual Energy Outlook 2013, Oil and Gas Supply Module*, (May 2013), http://www.eia.gov/forecasts/aeo/assumptions/pdf/oilgas.pdf.
[17] University of Southern California, USC Price School of Public Policy, *The Monterey Shale and California's Economic Future*, (March 2013), http://gen.usc.edu/assets/001/84955.pdf.

3 MONTEREY SHALE DISTRIBUTION AND HISTORICAL PRODUCTION

Figure 7 illustrates the distribution of potential tight oil production from the Monterey shale as considered in the EIA/INTEK report. Potential lies primarily in the San Joaquin, Santa Maria, and Ventura Basins, principally in Santa Barbara and Kern counties, although limited potential may exist in San Luis Obispo, Ventura, Kings, and Los Angeles counties (see Figure 1). The map indicates Occidental Petroleum Corporation leases (873,000 acres); an additional 248,500 acres owned by other companies comprise the remainder of the play.

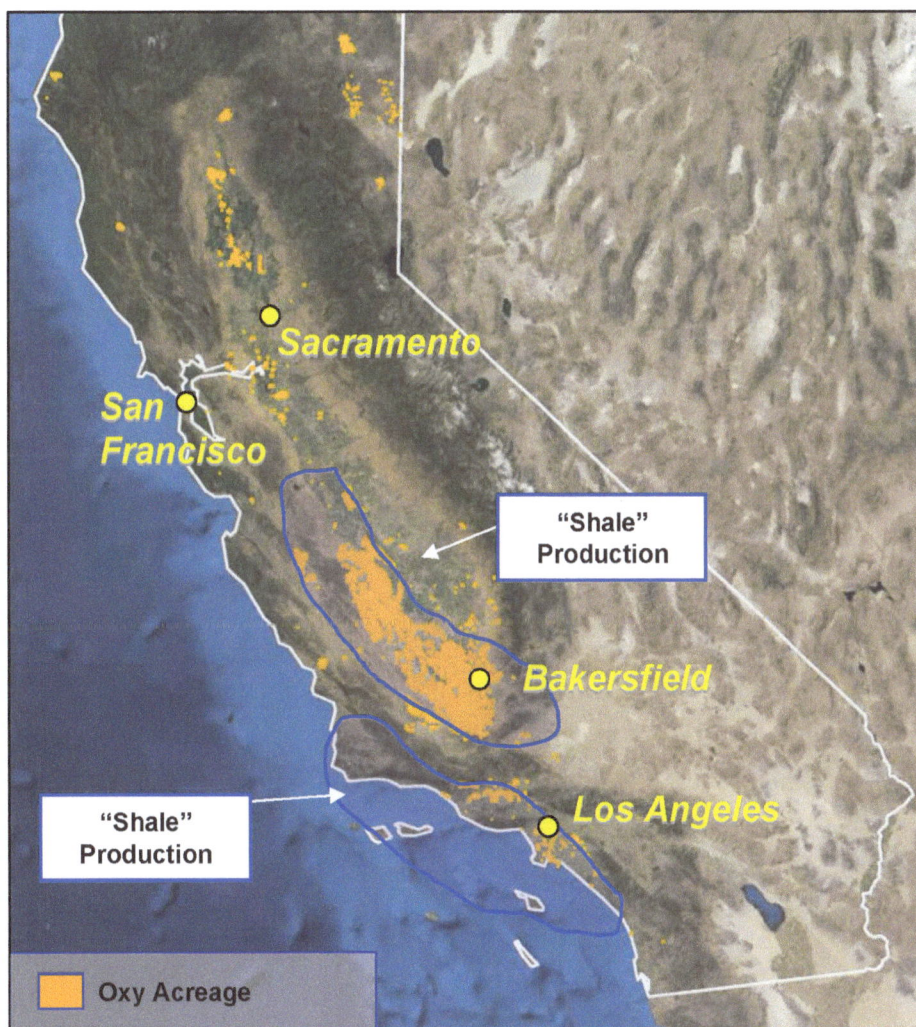

Figure 7. Distribution of potential tight oil production from the Monterey shale as considered in the EIA/INTEK report (2011).[18]

Orange areas are leases held by Occidental Petroleum Corporation, the largest leaseholder. The blue outlines are the general boundaries of the Monterey tight oil play.

---

[18] INTEK, Inc., *Review of Emerging Resources: U.S. Shale Gas and Shale Oil Plays*, December 2010, in U.S. Energy Information Administration, *Review of Emerging Resources: U.S. Shale Gas and Shale Oil Plays*, July 2011, http://www.eia.gov/analysis/studies/usshalegas/.

The Monterey Formation is complex, both stratigraphically and structurally, given the active tectonic regime in which it was deposited. Rock types are heterogeneous: although predominantly diatomaceous shale, they include silt and sand, and are variable in mineralogy and degree of diagenesis. The Monterey has long produced oil, much of it from the Stevens Sand—a conventional reservoir in the San Joaquin Basin—as well as from shale in the San Joaquin Basin and onshore and offshore in the Santa Maria and Ventura Basins of Santa Barbara County (see Section 1 for a full discussion of the formation's geological characteristics).

3.1   Overall Drilling and Production

Figure 8 illustrates Monterey oil production over the past three decades. Since 1977, the Monterey has produced 1.05 billion barrels of oil and 4.05 trillion cubic feet of gas, or nearly 10 percent of California's post-1977 oil production and more than 20 percent of its gas production.

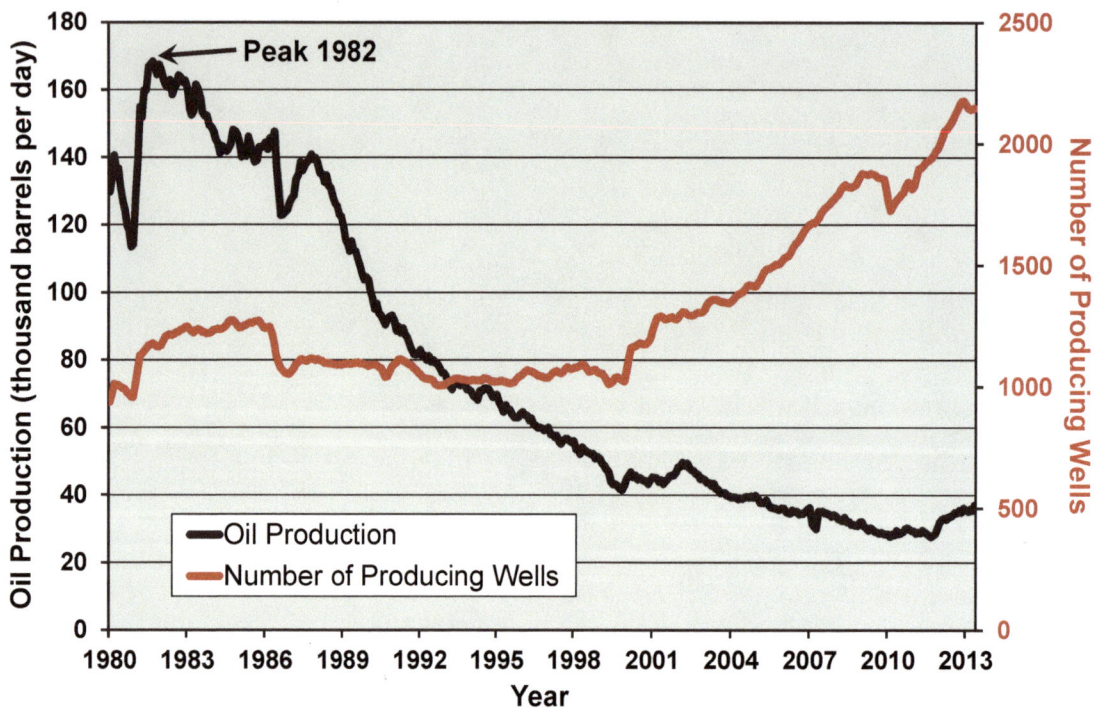

Figure 8. Monterey Formation oil production and number of producing wells, 1980 through May 2013.[19]

Production has declined from an average of over 140 barrels per day per well in the early 1980s to the current average of 17 barrels per day. Note that this includes production from the Stevens Sand member of the Monterey, which is a conventional reservoir.

---

[19] Data from DI Desktop (Drillinginfo), current through May 2013.

Production of oil and gas from the Monterey Formation has required the drilling of nearly 5,000 wells, of which 2,137 (42%) are currently non-productive (Figure 9). This is likely a minimum estimate, as some wells in fields that produce from the Monterey are not specifically linked to Monterey reservoirs in the Drillinginfo production database, and hence would not be included in this total. In some fields Monterey production is also commingled with production from overlying—and in some cases underlying— reservoirs.

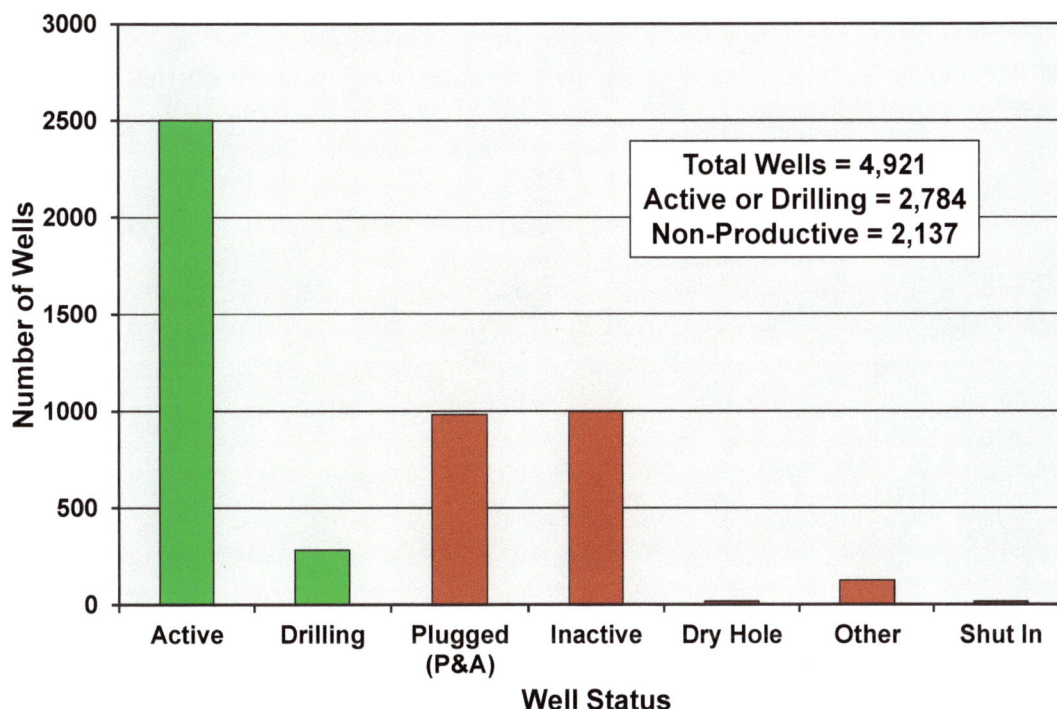

Total Wells = 4,921
Active or Drilling = 2,784
Non-Productive = 2,137

Figure 9. Status of Monterey oil and gas wells drilled from 1980 through May 2013.[20]

This is likely a minimum estimate of the number of wells as a considerable number of wells in the Drillinginfo database are not linked to a specific reservoir. Note that this includes wells for the Stevens Sand member of the Monterey, which is a conventional reservoir. P&A = plugged and abandoned.

[20] Data from DI Desktop (Drillinginfo), current through May 2013.

The geographic distribution of these wells is illustrated in Figure 10. The majority of the drilling has occurred in the southern San Joaquin Basin, mainly in Kern County. There are, however, approximately 500 wells drilled in the Santa Maria and Ventura Basins of Santa Barbara County, including significant offshore production from the Holly Platform in the South Elwood Field located in State waters.

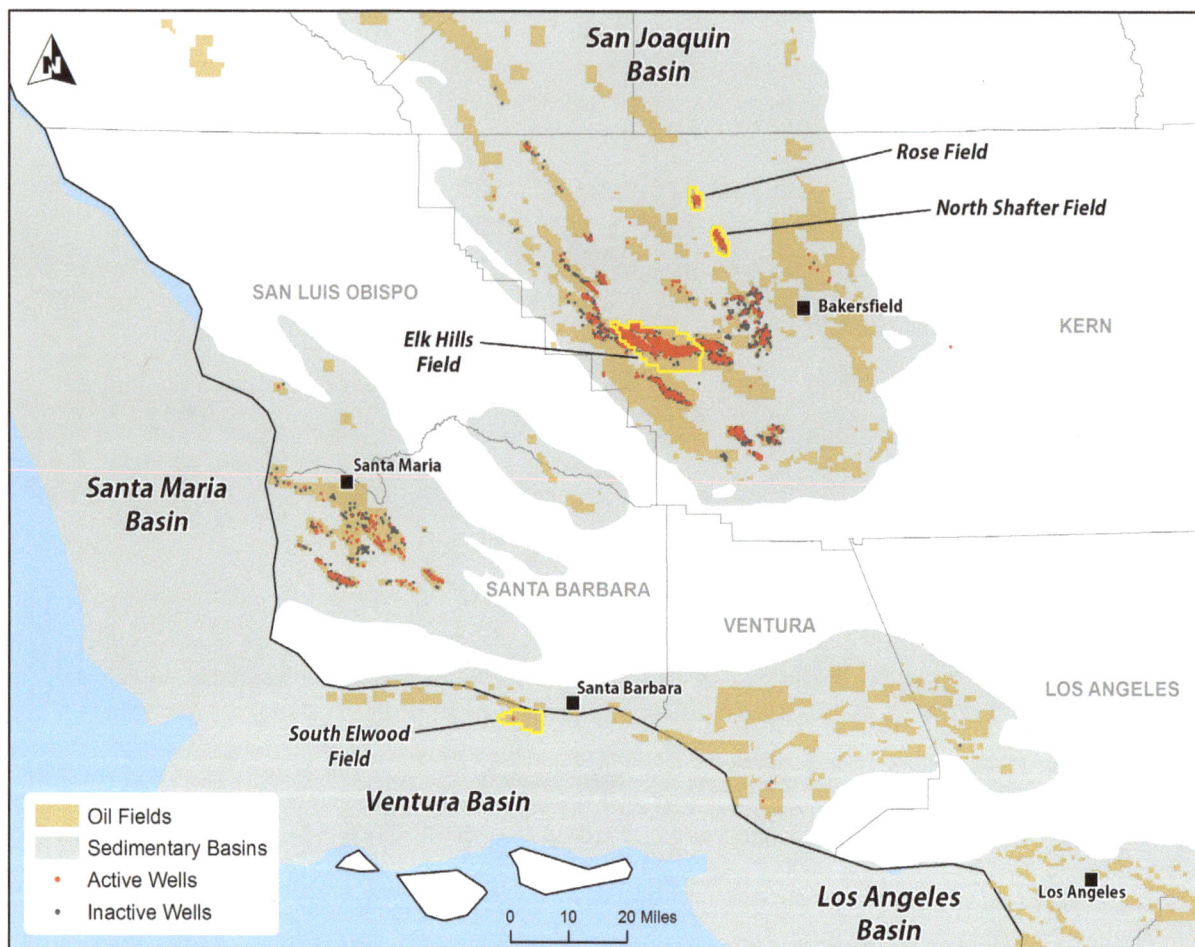

Figure 10. Distribution of oil fields, including wells producing from the Monterey Formation as of May 2013.[21]

The fields mentioned in this report are outlined in yellow. Sedimentary basins that include Monterey Formation rock are also shown.

---

[21] Well location data from DI Desktop (Drillinginfo), current through May 2013. Field location data from DOGGR, current through October 2013.

3.2 Production by Subdivision

The Monterey Formation in the San Joaquin Basin is subdivided into several members as illustrated in Figure 11. Significant production has historically been obtained from the Stevens Sand—essentially a conventional reservoir charged by adjacent Monterey shale source rocks—as well as from the Antelope Shale, McLure Shale, Reef Ridge Shale, and Devilwater Shale. In the Santa Maria and Ventura Basins of Santa Barbara County the Monterey Formation is not subdivided except geographically, into onshore and offshore components.

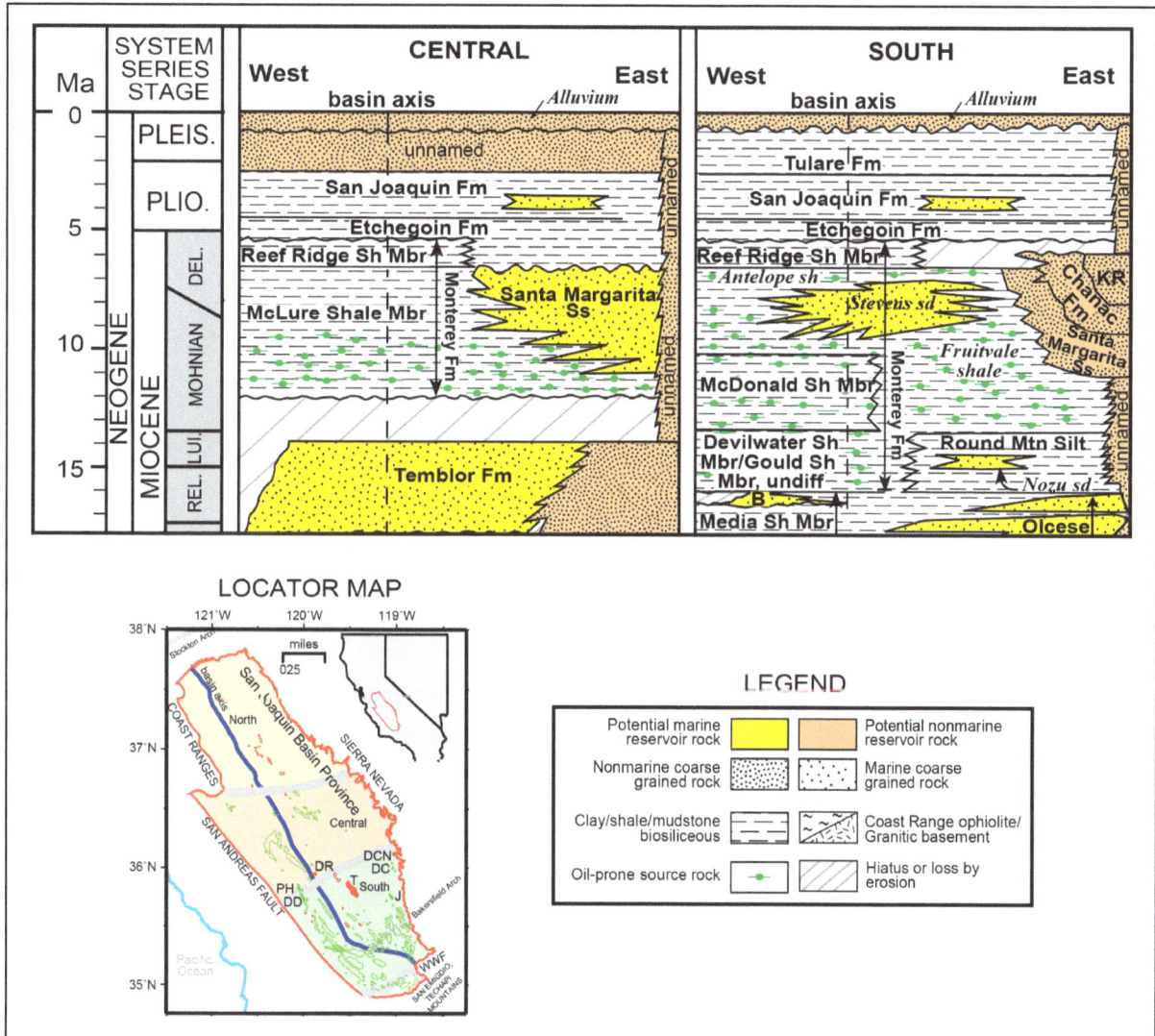

Figure 11. Stratigraphic nomenclature of the Monterey Formation and adjacent strata in the central and southern San Joaquin Basin.[22]

Note the vertical arrows indicating the strata that belong to the Monterey Formation.

[22] A. H. Scheirer, ed., *Petroleum Systems and Geologic Assessment of Oil and Gas in the San Joaquin Basin Province, California*, U.S. Geological Survey Professional Paper 1713, (2007), http://pubs.usgs.gov/pp/pp1713/.

Oil production from these subdivisions of the Monterey Formation is illustrated in Figure 12. As can be seen, production is dominated by the conventional Stevens Sand, which is being produced from several fields in the San Joaquin Basin. Another major producer of oil from the Monterey Formation has been the South Elwood Offshore Field ("Offshore Santa Barbara" in the Figure) via the Holly Platform west of Santa Barbara. As of mid-2013, total oil production from the entire Monterey Formation was approximately 36,000 barrels per day (six-month moving average).

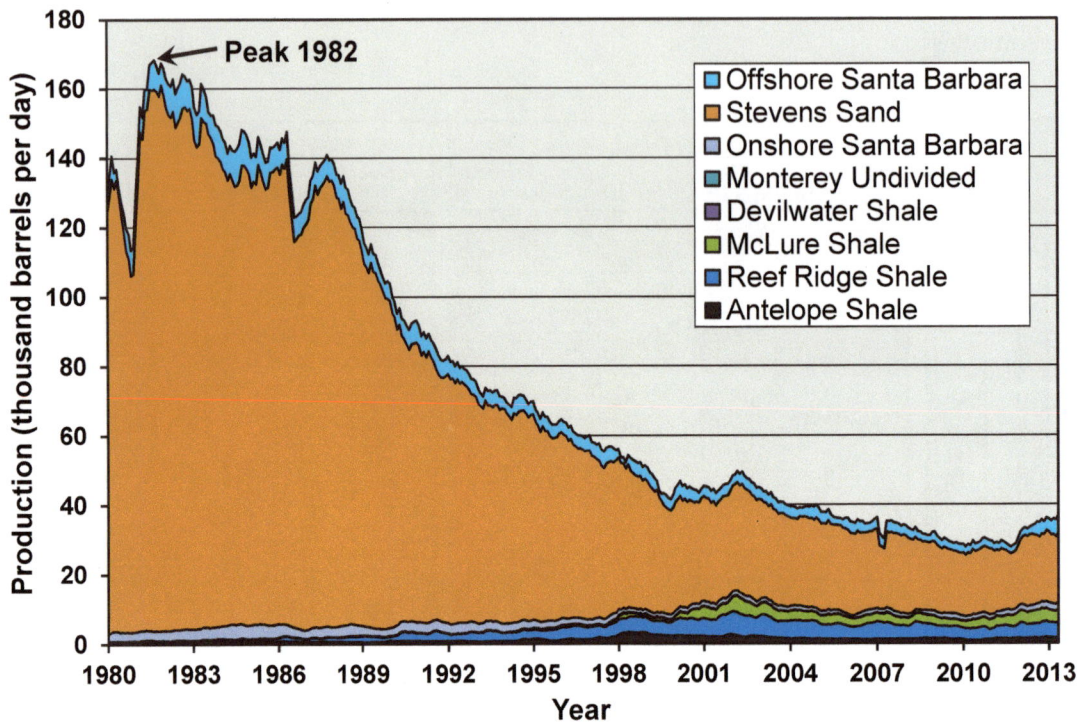

Figure 12. Oil production from subdivisions of the Monterey Formation, 1980 through May 2013.[23]

"Offshore Santa Barbara" and "Onshore Santa Barbara" are contained in the Santa Maria and Ventura Basins; all other subdivisions are contained in the San Joaquin Basin.

[23] Data from DI Desktop (Drillinginfo), current through May 2013.

The oil produced from the Stevens Sand Reservoir and the South Elwood Field is migrated oil discovered through typical exploration techniques and extracted mostly through conventional techniques (albeit possibly with some cyclic steam injection). Thus, neither the Stevens Sand reservoir nor the South Elwood Field are what the EIA/INTEK report was referring to in terms of potential new Monterey tight oil production. The Stevens Sand is a conventional reservoir, not a tight oil reservoir, and it is in decline.

Offshore oil exploration and drilling has long been subject to a moratorium; Venoco, the owner of the Holly Platform (pictured in Figure 13), recently abandoned a plan to expand drilling from it,[24] and in 2011 a 5-year ban on new offshore drilling was imposed on the entire U.S. West Coast.[25] Resistance to offshore drilling on the Pacific Coast stemmed from the 1969 Santa Barbara spill, the third worst in U.S. history after the Macondo (Deepwater Horizon) and Exxon Valdez spills.

Directional drilling of offshore oil reservoirs from onshore sites has been considered as a way of circumventing restrictions on offshore drilling. However, this practice would need to pass many legal and public opposition hurdles before being implemented to any significant degree.

Thus it is the onshore portions of the Santa Maria and San Joaquin Basins—excluding conventional reservoirs like the Stevens Sand—that are the focus of the EIA/INTEK report and therefore this study. (There is currently no significant onshore oil production from the Monterey in the Ventura or Los Angeles Basins, so they are not included in this study.)

Figure 13. Holly Platform producing from the Monterey Formation in the South Elwood Field.
The Holly Platform, constructed in 1966 in 211 feet of water, is owned by Venoco and produces a significant portion of Monterey Formation oil in Santa Barbara County from a naturally fractured shale reservoir.[26] See Figure 10 for location.

---

[24] Pacific Coast Business Times, "Venoco drops drilling plans," November 17, 2010, http://www.pacbiztimes.com/2010/11/17/venoco-drops-ellwood-drilling-plans/.
[25] Committee on Natural Resources, U.S. House of Representatives, "Obama Administration Imposes Five-Year Drilling Ban on Majority of Offshore Areas," November 8, 2011, http://naturalresources.house.gov/news/documentsingle.aspx?DocumentID=267985.
[26] Photo by Andy Rusch, licensed under Create Commons, http://commons.wikimedia.org/wiki/File:Platform_Holly,_Santa_Barbara.jpg.

Average oil production per well in each of the oil-bearing subdivisions of the onshore portions of the Monterey Formation is illustrated in Figure 14. Production from the Stevens Sand, Antelope Shale, and onshore Monterey in the Santa Maria Basin of Santa Barbara County ("Onshore Santa Barbara" in the chart) currently averages between 8 and 18 barrels per day per well. The McLure Shale and Reef Ridge Shale of the San Joaquin Basin average nearly 40 barrels per day per well, roughly four times the overall California average. The McLure Shale in the Rose and North Shafter Fields of Kern County is a relatively new development that is accessed by horizontal and vertical wells stimulated by fracking, and is discussed in more depth below.

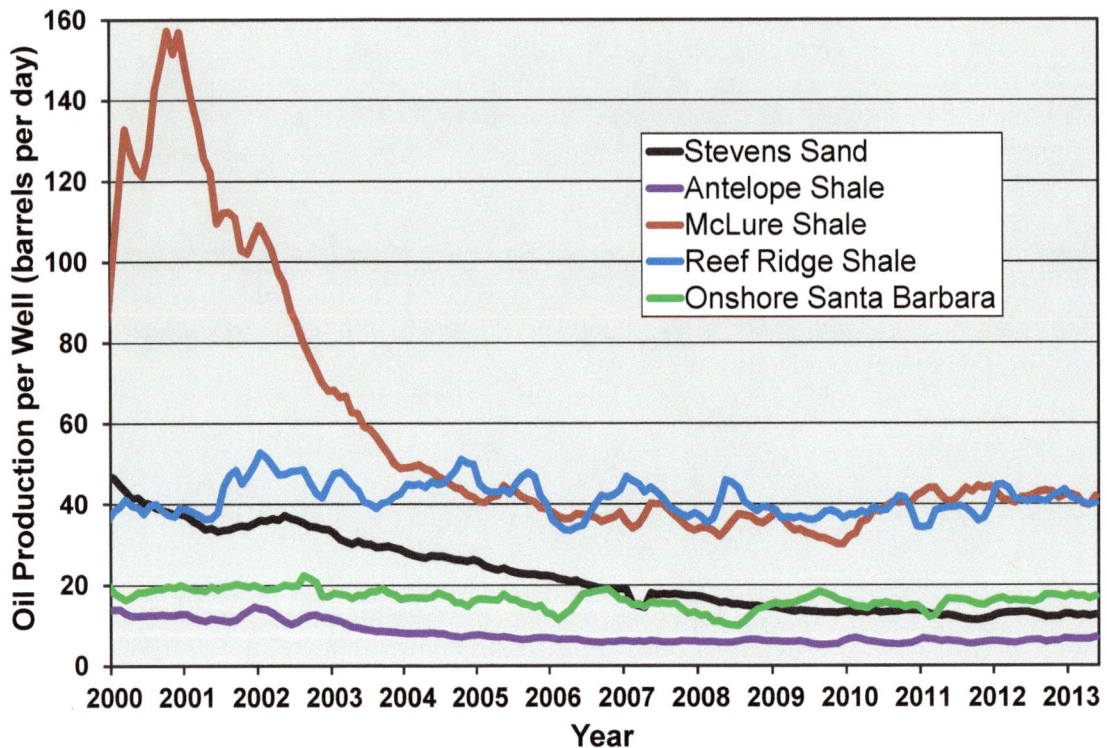

Figure 14. Average onshore oil production per well from subdivisions of the Monterey Formation, 2000 through May 2013.[27]

"Onshore Santa Barbara" is contained in the Santa Maria Basin of Santa Barbara County and all other subdivisions are contained in the San Joaquin Basin. There is currently no significant onshore production from the Ventura or Los Angeles Basins.

---

[27] Data from DI Desktop (Drillinginfo), current through May 2013.

The portion of oil production from the Monterey Formation that relates to shale reservoirs potentially accessible to development is illustrated in Figure 15. (Figure 15 shows the strata shown in Figure 12 minus the Stevens Sand reservoir and Offshore Santa Barbara.) These reservoirs saw a collective production peak in 2002 at a little over 15,000 barrels per day, or about three percent of total California production. Currently they produce about 11,500 barrels per day (six-month moving average), or 32% of all oil from the Monterey Formation.

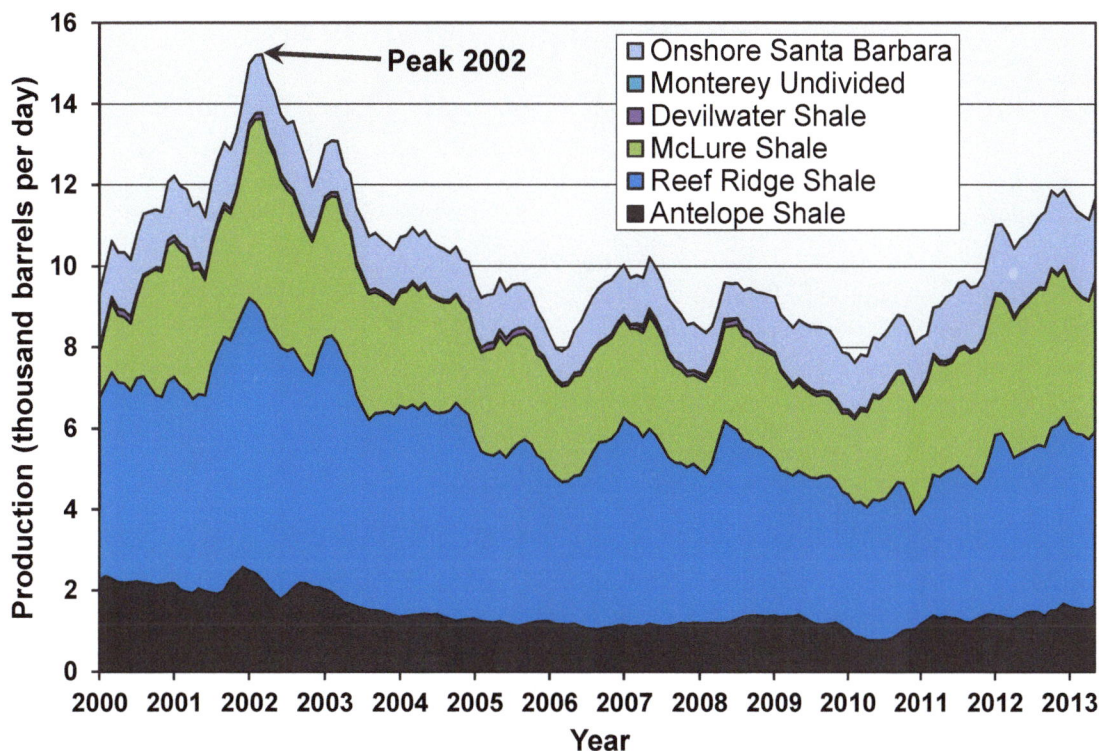

Figure 15. Oil production from shale reservoirs in the Monterey Formation, 2000 through May 2013.[28]

"Onshore Santa Barbara" is contained in the Santa Maria Basin of Santa Barbara County; all other subdivisions are contained in the San Joaquin Basin of Kern County.

Overall oil production in California has been in decline for decades. The question is whether the Monterey Formation can produce a new tight oil production boom with the application of advanced production techniques. The next section of this report investigates this question by assessing the geological characteristics and production history of the Monterey shale and comparing them to other tight oil plays. It also examines the assumptions used by the EIA/INTEK and USC studies in developing estimates of recoverable oil and economic benefits.

---

[28] Data from DI Desktop (Drillinginfo), current through May 2013.

4 Geological Characteristics of the Monterey

The geological characteristics of the Monterey Formation and the history of oil production there are analyzed below to evaluate the credibility of the assumptions in the EIA/INTEK report, and of the estimated potential of tight oil production from the Monterey.

4.1 General Comparison to the Bakken and Eagle Ford

Much of the enthusiasm about developing tight oil in the Monterey comes from the success of tight oil production in the Bakken and Eagle Ford plays of North Dakota and southern Texas, respectively. Geologically, however, the Monterey is very different from these two formations, and presents very different challenges to exploration and drilling.

- The Bakken and Eagle Ford plays are relatively predictable, with target strata less than a few hundred feet in thickness and flat-lying to gently dipping. The Monterey is a much thicker and much more complex shale formation, with target strata up to 2,000 or more feet in thickness in the San Joaquin Basin and even thicker in the Santa Maria Basin and offshore in the Ventura Basin. Moreover, target strata depths can range from outcrops to 18,000 feet within a span of forty miles or less.

- The Bakken play is spread over a potentially productive area of as much as 20,000 square miles, and the Eagle Ford play covers roughly 8,000 square miles[29]; horizontal laterals as long as 10,000 feet are routinely drilled in the Bakken.[30] The Monterey tight oil play encompasses only 1,752 square miles; production from the Monterey has to date been largely from areally restricted structural and stratigraphic traps.

- The Bakken and Eagle Ford shales are approximately 360 and 90 million years old, respectively, and were deposited on a relatively stable platform; as a result they are thin and widespread, and hence are relatively predictable. The Monterey is 6-16 million years old and was deposited rapidly in an active tectonic regime; as a result it is thick, of limited areal extent, and was deformed after deposition by movement on the San Andreas Fault and other structures, and hence is much less predictable.

---

[29] U.S. Energy Information Administration, *Assumptions to the Annual Energy Outlook 2013, Oil and Gas Supply Module*, (May 2013), http://www.eia.gov/forecasts/aeo/assumptions/pdf/oilgas.pdf.
[30] North Dakota Industrial Commission GIS Map Viewer, https://www.dmr.nd.gov/OaGIMS/viewer.htm.

4.2   Depth, Thickness and Areal Extent

The distribution of the McLure Shale and Antelope Shale source rocks of the Monterey in the San Joaquin Basin is illustrated in Figure 16; these are also the major petroleum system source rocks that have charged many conventional reservoirs. Burial depths range from less than 2,000 feet to more than 18,000 feet over distances of 40 miles or less. At depths of greater than about 11,000 to 14,000 feet, these rocks are still in the oil and gas generation windows (i.e., they are still generating and expelling oil and gas). They range up to 2,000 or more feet in thickness in the San Joaquin Basin, and are even thicker in the Santa Maria Basin and offshore in the Ventura Basin. They are also complexly folded and faulted, at macro-, meso-, and micro-scales.

The cross sections in Figure 17 illustrate folding in the southern and central San Joaquin Basin, and Figure 18 illustrates deformation at the meso-scale along the beach near Vandenberg Air Force base in the Santa Maria Basin. One might argue that this bodes well for permeability from naturally induced fractures—others might suggest that this has expedited the migration of oil from these source rocks to overlying structural and stratigraphic traps. Included in the latter are companies like Chevron, who stated:

> 'Based on our drilling results, our view is that the oil has migrated out of the formation and is now found in pockets outside of the Monterey shale,' said Kurt Glaubitz, a spokesman for San Ramon, California-based Chevron, the second-biggest U.S. oil producer.[31]

This level of complexity is atypical of productive tight oil plays like the Bakken and the Eagle Ford. There is no question that the Monterey has long produced oil but, as illustrated in Figure 12 and Figure 15, the vast majority of that is from conventional wells producing oil that has migrated significant distances from source rocks. The Monterey's best years may already be behind it.

---

[31] Chris Reed, "Wall Street doubts CA shale hype – but not Occidental," *CalWatchdog*, April 11, 2013, http://calwatchdog.com/2013/04/11/wall-street-doubts-ca-shale-hype-but-not-occidental/.

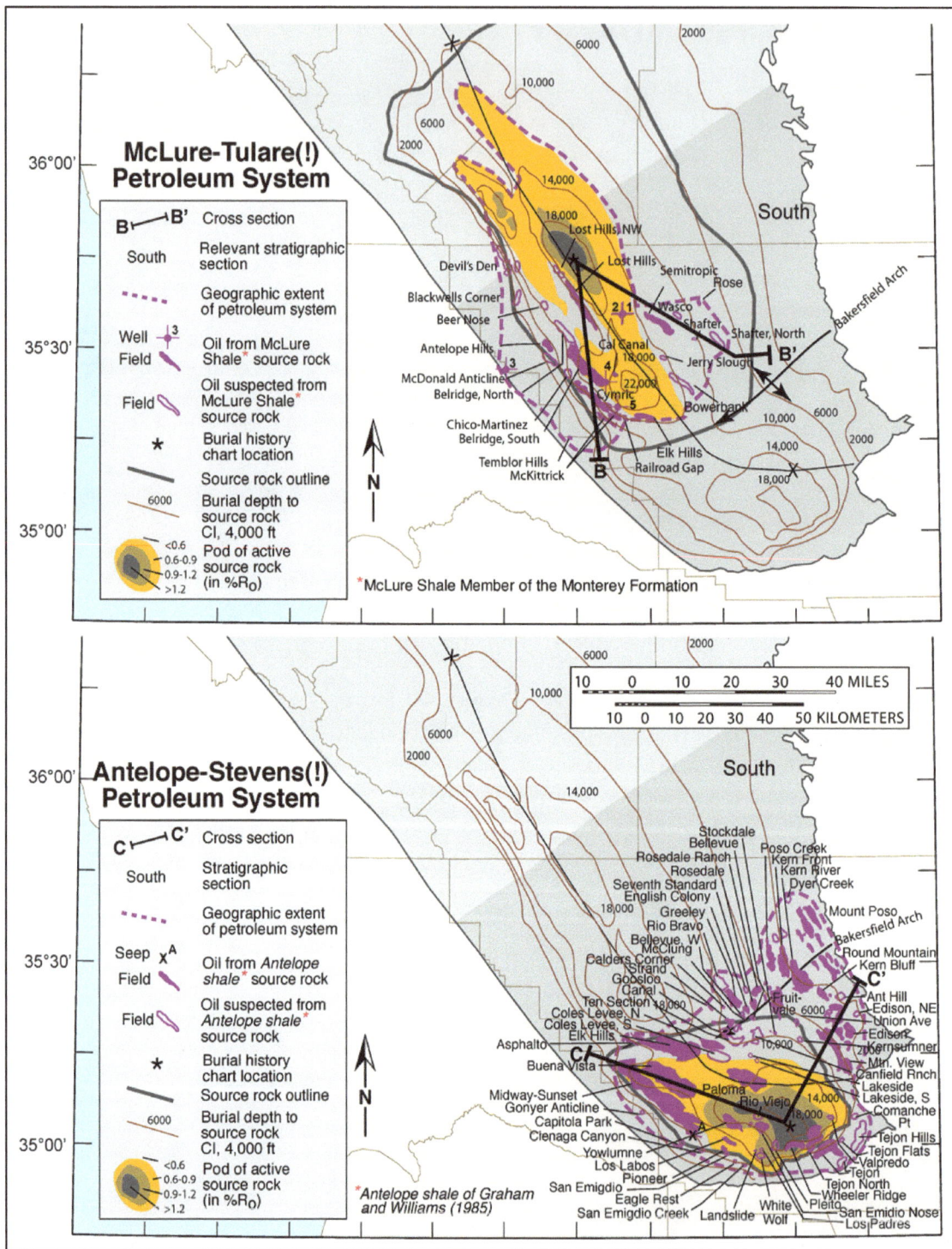

Figure 16. Distribution and depth of the McLure Shale (top) and Antelope Shale (bottom) source rocks of the Monterey in the central and southern San Joaquin Basin.[32]

Purple areas represent fields charged with oil that migrated from these source rocks. Orange and grey colored areas are within the oil and gas generation window. The location of the cross sections illustrated in Figure 17 are indicated.

[32] A. H. Scheirer, ed., *Petroleum Systems and Geologic Assessment of Oil and Gas in the San Joaquin Basin Province, California*, U.S. Geological Survey Professional Paper 1713, (2007), http://pubs.usgs.gov/pp/pp1713/.

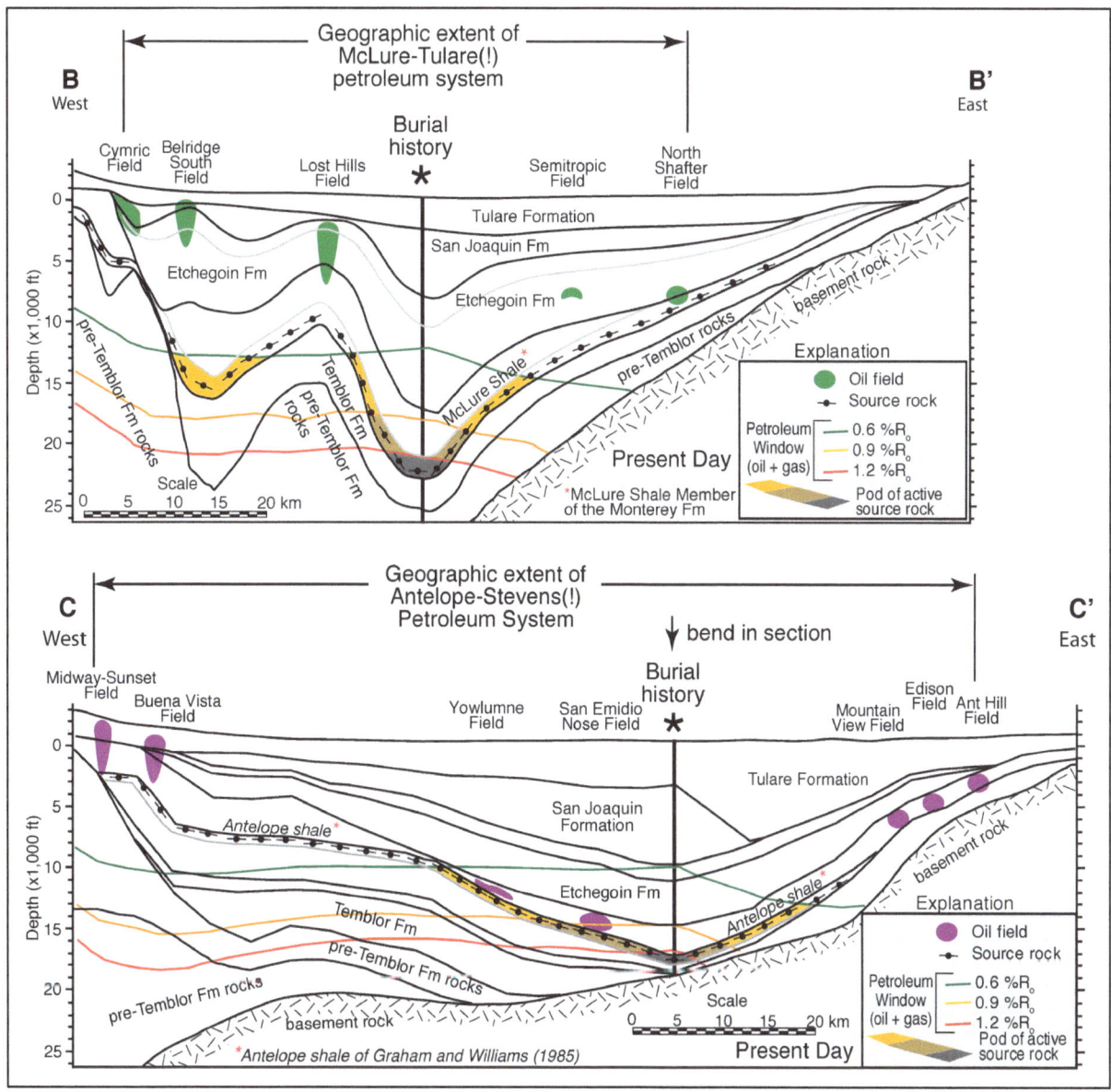

Figure 17. Cross sections illustrating the distribution and depth of the McLure Shale (top) and Antelope Shale (bottom) source rocks of the Monterey in the central and southern San Joaquin Basin.[33]

Green and purple areas represent fields charged with oil that migrated from these source rocks. Orange and grey colored areas are within the oil and gas generation window. See Figure 16 for the locations of these cross sections.

[33] A. H. Scheirer, ed., *Petroleum Systems and Geologic Assessment of Oil and Gas in the San Joaquin Basin Province, California*, U.S. Geological Survey Professional Paper 1713, (2007), http://pubs.usgs.gov/pp/pp1713/.

Figure 18. Outcrop of Monterey Formation strata along the coast at Vandenberg Air Force Base illustrating complex meso-scale deformation.[34]

[34] L.S. Durham, "Monterey Shale Continues to Tempt and Tease," *AAPG Explorer*, February 2013, https://www.aapg.org/explorer/2013/02feb/monterey_shale0213.cfm.

4.3   Distribution of Production Potential

As noted earlier, the Monterey Formation is a very important source rock and oil expelled from it has charged many of the most productive conventional oil fields in the San Joaquin Basin (and to a lesser extent conventional fields in the Santa Maria and Ventura Basins). The U.S. Geological Survey estimates that 422 oil fields, oil seeps, and isolated oil wells have been charged with oil derived from the Monterey in the San Joaquin Basin.[35] This oil has been trapped in conventional structural, stratigraphic and fault traps discovered primarily by seismic- and surface-mapping. They are not ubiquitous and in fact cover limited parts of the central and southern San Joaquin Basin, as can be seen on the maps in Figure 16 and on the cross sections in Figure 17.

The current oil generation window in the San Joaquin Basin lies at depths of between 11,000 and 18,000 feet (Figure 16 and Figure 17). The extent of source rocks containing oil that has not migrated, comparable to other tight oil plays, depends on the existence of source rocks that have passed through the oil generation window and subsequently been uplifted to the target depths in the EIA/INTEK report (8,000-14,000 feet). The areal extent of such rocks, if they exist, is likely to be far smaller than the region outlined in the EIA/INTEK report (Figure 7).

In the Santa Maria and Ventura Basins, Monterey production is almost exclusively from migrated oil in structural, stratigraphic, and faulted traps. Productive areas are not ubiquitous (see distribution of Monterey wells in Figure 10). This is particularly well illustrated by production from the Holly Platform off the coast of Santa Barbara, where wells produce at high rates from a fractured shale reservoir in the South Elwood Field of the Monterey, yet are surrounded by a much larger region of low productivity or dry wells. The existence of widespread mature source rocks containing oil that has not migrated at accessible depths and with available surface access in the Santa Maria and Ventura Basins of Santa Barbara County is speculative at best.

---

[35] A. H. Scheirer, ed., *Petroleum Systems and Geologic Assessment of Oil and Gas in the San Joaquin Basin Province, California*, U.S. Geological Survey Professional Paper 1713, (2007), http://pubs.usgs.gov/pp/pp1713/, Appendix 8.1.

4.4　Well Productivity and Oil Recovery

The economics of an oil well depend on well cost, well productivity, and how much oil the well recovers over its lifetime (ultimate recovery), as well as the cost of infrastructure and the land on which it is drilled. Well productivity generally declines over time, with the highest output immediately after it is drilled—its **initial productivity** (IP), defined as the average daily production in the highest month of production from a well (which is normally the first or second month after a well is drilled). Enhanced oil recovery techniques such as water-flooding, CO_2 flooding, or cyclic-steam injection are utilized to help maintain oil production over time to increase the **cumulative recovery** of wells in suitable reservoirs.

The EIA/INTEK report utilizes assumptions of average initial productivity and well production declines from the Monterey Formation in the Elk Hills Field. The USC study utilizes the oil production profile of wells in the Rose and North Shafter fields as prototypes for future Monterey tight oil production. To assess the validity of these assumptions, this section of the report reviews the production history and geological characteristics of the Elk Hills, Rose, and North Shafter Fields (see Figure 10 for location), followed by an analysis of the initial productivity and cumulative recovery of all shale wells in the San Joaquin and Santa Maria Basins (which is where virtually all Monterey oil production has taken place).

4.4.1　Monterey Shale Prototype: Elk Hills Field

The EIA/INTEK report provides the diagram in Figure 19, attributed to Occidental Petroleum Corp. (Oxy), as an example of the expected IP and typical production decline curves for Monterey shale wells. One of these curves is indicated as being from a typical "Elk Hills Area 'Shale' Vertical Well" which has an IP of 800 barrels of oil equivalent (BOE) per day.[36] Typical horizontal and vertical shale wells for non-specific locations within the Monterey tight oil play are said to have IPs of 500 and 200 barrels of oil equivalent per day, respectively.[37]

---

[36] Oil equivalent is the sum of oil output and gas output converted into the energy equivalent of oil. In the case of the Elk Hills Field, roughly half of the energy output from a well comes from natural gas.

[37] INTEK, Inc., *Review of Emerging Resources: U.S. Shale Gas and Shale Oil Plays*, December 2010, in U.S. Energy Information Administration, *Review of Emerging Resources: U.S. Shale Gas and Shale Oil Plays*, July 2011, http://www.eia.gov/analysis/studies/usshalegas/.

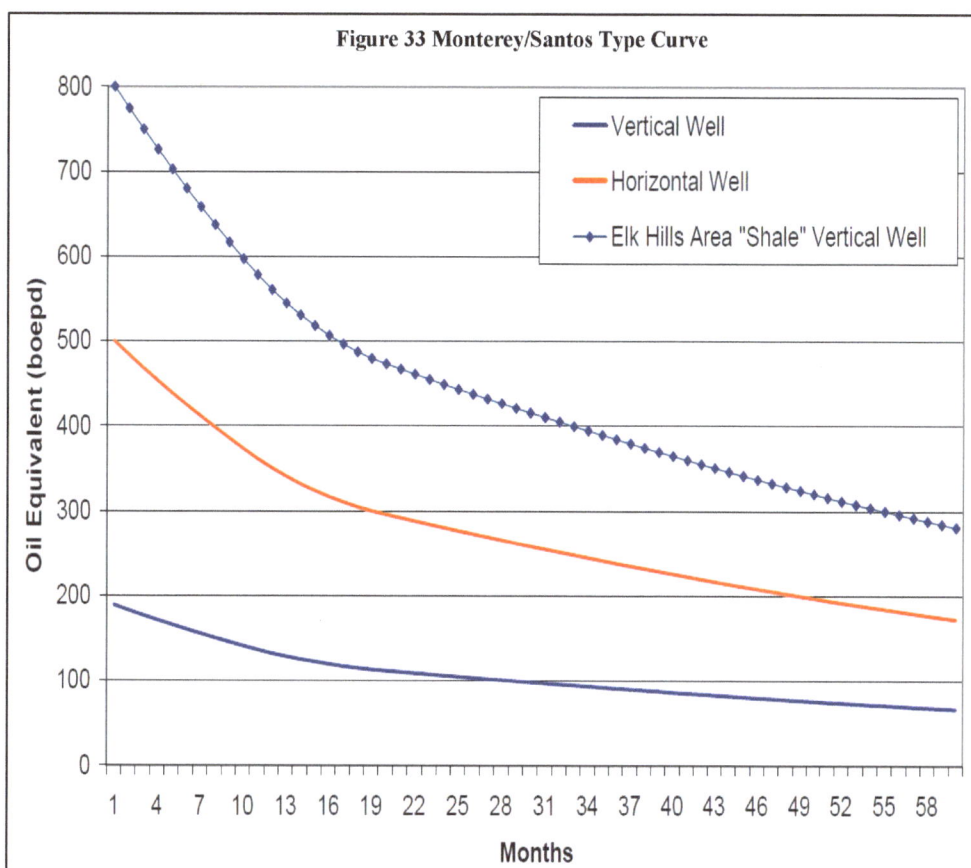

Figure 33 Monterey/Santos Type Curve

Figure 19. Typical type decline curves for Monterey shale wells provided in the EIA/INTEK report.[38]

Figure is presented as shown in the EIA/INTEK report, in which the curves are attributed to Occidental Petroleum Corporation.

Two things strongly call into question these so-called "typical" shale well decline curves presented by EIA/INTEK. The first is that the Elk Hills Field primarily produces from the Stevens Sand member of the Monterey Formation, a conventional reservoir charged with migrated oil from Monterey source rocks. Thus it is in no way "typical" of what to expect from a tight oil well, which would produce from non-migrated oil in source rock as done in other tight oil plays. Secondly, of the 1,212 wells drilled by Occidental over the last six years in the Elk Hills Field, only six had IPs of over 800 BOE/day (Figure 20). In fact, the average IPs of all directional wells (which include horizontal wells) and vertical wells drilled by Occidental in the Elk Hills Field in the past six years are 108 and 96 BOE/day, respectively. On an "oil only" basis (i.e., excluding the energy in the associated gas produced) Occidental's Elk Hills wells averaged only 46 barrels per day for both directional and vertical wells (Figure 21).

Even if one accepts the proposition that production from the Stevens Sand in the Elk Hills Field is representative of what to expect from typical Monterey shale wells, the shale well decline curves in the EIA/INTEK report (Figure 19) appear to be highly overstated when compared to actual production data.

---

[38] INTEK, Inc., *Review of Emerging Resources: U.S. Shale Gas and Shale Oil Plays*, December 2010, in U.S. Energy Information Administration, *Review of Emerging Resources: U.S. Shale Gas and Shale Oil Plays*, July 2011, http://www.eia.gov/analysis/studies/usshalegas/. This is Figure 33 in the report.

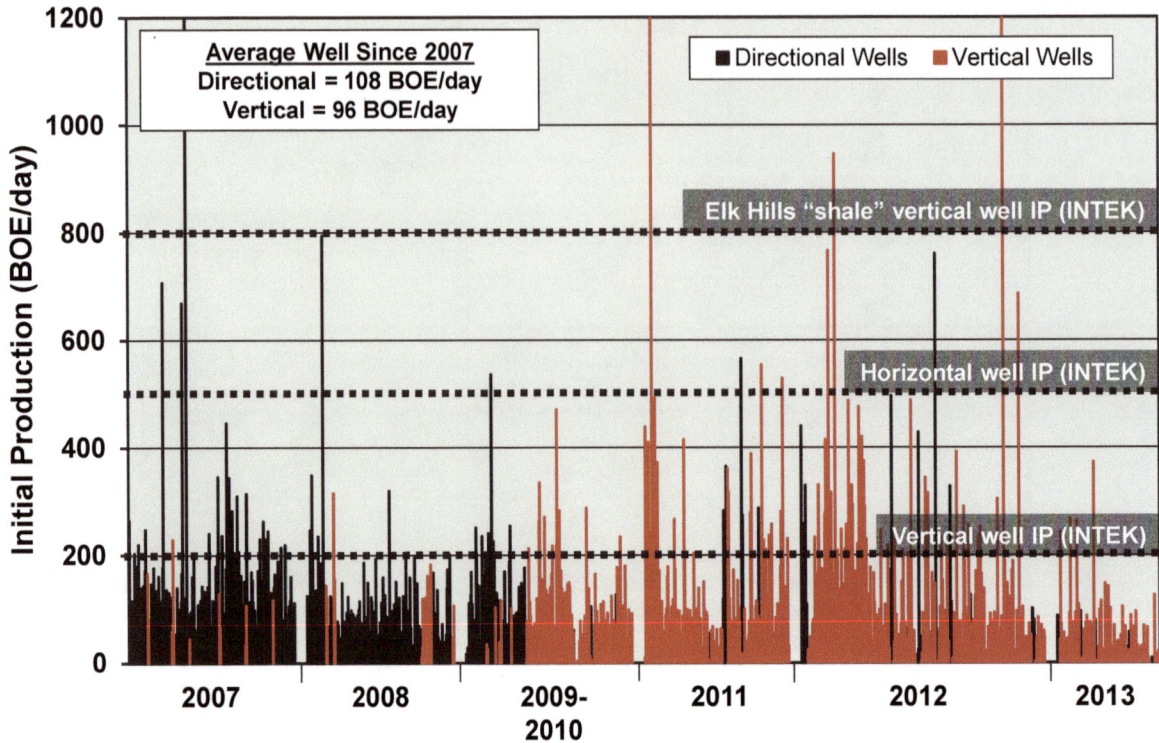

Figure 20. Initial productivity for all Elks Hills Field wells drilled by Occidental, grouped by year of first production, 2007 through June 2013.[39]

2,012 wells in total; production is in barrels of oil equivalent (BOE) per day. The IPs for "typical" wells as claimed in the EIA/INTEK report (see Figure 19) are indicated. The average IPs of all directional and vertical wells drilled in the past six years are 108 and 96 BOE/day, respectively.

---

[39] Data from DI Desktop (Drillinginfo), current through June 2013.

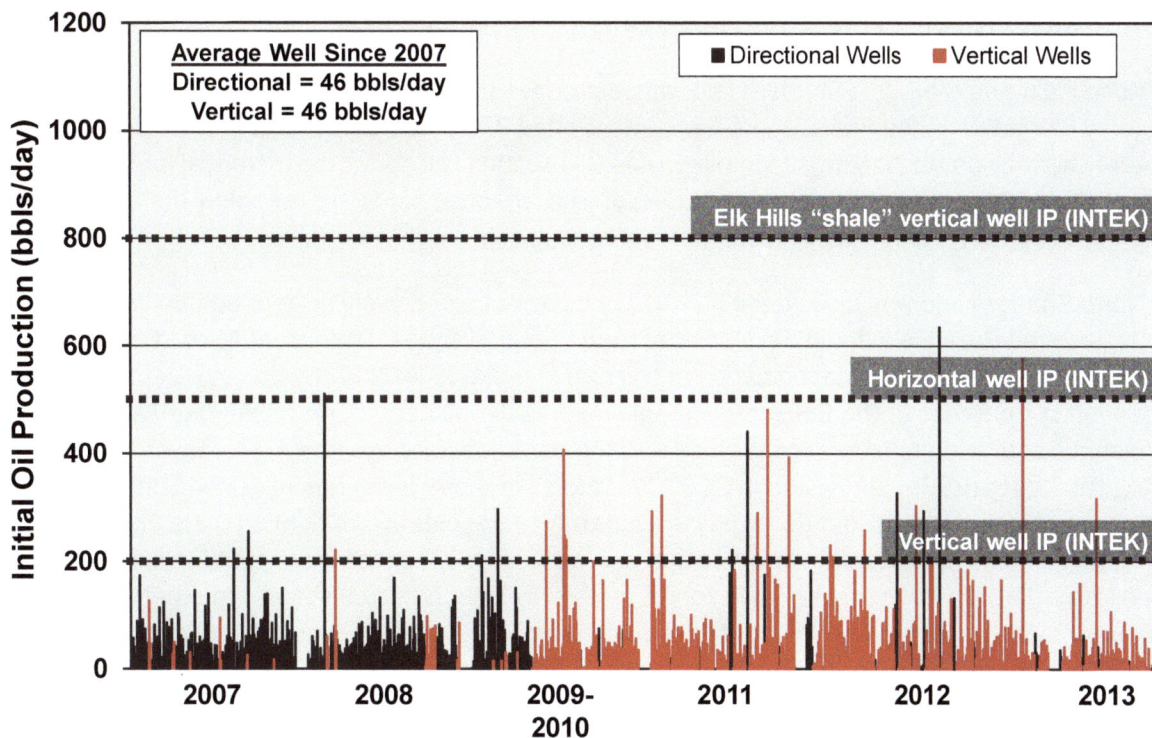

Figure 21. Initial productivity on an "oil only" basis for all Elk Hills Field wells drilled by Occidental, grouped by year of first production, 2007 through June 2013.[40]

2,012 wells in total. The IPs for "typical" wells as claimed in the EIA/INTEK report (see Figure 19) are indicated (but are on a barrels of oil equivalent basis; see Figure 20).

---

[40] Data from DI Desktop (Drillinginfo), current through June 2013.

4.4.2 Monterey Shale Prototype: Rose and North Shafter Fields

The Rose Field and the North Shafter Field, which produce from the McLure Shale member of the Monterey Formation in the San Joaquin Basin, were cited in the USC study as examples of what Monterey tight oil production might look like.[41] As this section will show, the historical initial productivities (IP) and cumulative productivities of wells in these fields are far below the assumptions made in the EIA/INTEK report for the Monterey.

The North Shafter Field was discovered in 1983 but was not extensively developed until the mid- to late-1990s, with the Rose Field being developed around the turn of the 21st century. These fields produce from shale using both horizontal and vertical fracture stimulated wells. They do not resemble tight oil production in the sense of the Bakken and Eagle Ford as the oil appears to be migrated and trapped by a diagenetic seal (although the rock is shale and the permeabilities are low).[42] Moreover, the U.S. Geological Survey (USGS) suggested in 2003 that the oil in these fields has migrated some 15 miles from its source; if so, this also means they are not truly representative of tight oil plays elsewhere.[43] Finally, these fields are tightly constrained in a relatively small area, not widespread in sharp contrast to tight oil plays like the Bakken and Eagle Ford. They are well defined by a seismic anomaly as illustrated in Figure 22, and cover an area of about 20 square miles.

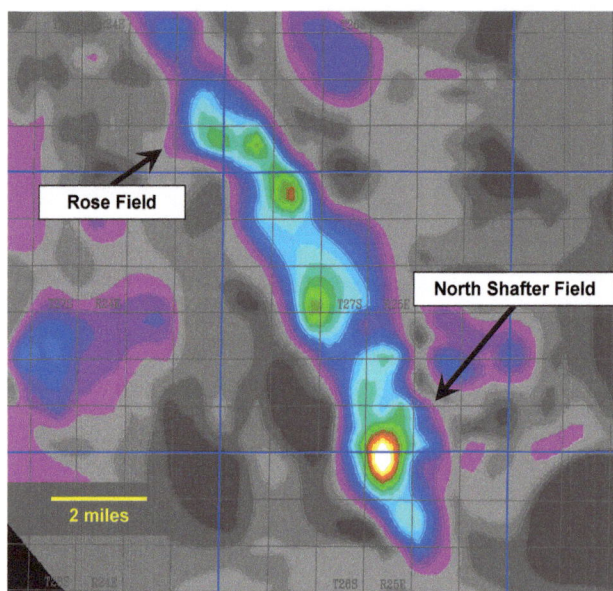

Figure 22. Seismic anomaly defining the Rose and North Shafter Fields.[44]

These fields are tightly constrained by non-productive rocks and occupy only about 20 square miles in area.

---

[41] University of Southern California, USC Price School of Public Policy, *The Monterey Shale and California's Economic Future*, (March 2013), http://gen.usc.edu/assets/001/84955.pdf; see Appendix K.

[42] See A. H. Scheirer, ed., *Petroleum Systems and Geologic Assessment of Oil and Gas in the San Joaquin Basin Province, California*, U.S. Geological Survey Professional Paper 1713, (2007), http://pubs.usgs.gov/pp/pp1713/, Appendix 8.1. Although the trap type is listed as "unknown" by the USGS, it appears to be a stratigraphic trap with an "Opal C/T" diagenetic phase providing the seal, and a reservoir of "fractured, porosity enhanced, oil-saturated quartz-phase rocks"; see A. Grau, R. Sterling, and R. Kidney, "Success! Using Seismic Attributes and Horizontal Drilling to Delineate and Exploit a Diagenetic Trap, Monterey Shale, San Joaquin Valley, California," (adapted from a poster presented at AAPG annual conference, Salt Lake City, May 2003); AAPG Datapages Search and Discovery Article #20011, Poster #2, http://www.searchanddiscovery.com/documents/2003/grau/index.htm.

[43] A. H. Scheirer, ed., *Petroleum Systems and Geologic Assessment of Oil and Gas in the San Joaquin Basin Province, California*, U.S. Geological Survey Professional Paper 1713, (2007), http://pubs.usgs.gov/pp/pp1713/.

[44] A. Grau, R. Sterling, and R. Kidney, "Success! Using Seismic Attributes and Horizontal Drilling to Delineate and Exploit a Diagenetic Trap, Monterey Shale, San Joaquin Valley, California," (adapted from a poster presented at AAPG annual conference, Salt Lake City, May 2003); AAPG Datapages Search and Discovery Article #20011, Poster #2, http://www.searchanddiscovery.com/documents/2003/grau/index.htm.

Figure 23 illustrates oil production and the number of producing wells over the last two decades for the Rose and North Shafter fields. Oil production peaked in 2002 at about 5,000 barrels per day and is now down 20 percent from that peak; production currently averages about 40 barrels per day per well. The distribution of wells for the two fields is illustrated in Figure 24 and an illustration of well footprint in the North Shafter Field is provided in Figure 25. These fields have produced a combined 15.2 million barrels of oil and 8 billion cubic feet of natural gas since their discovery.

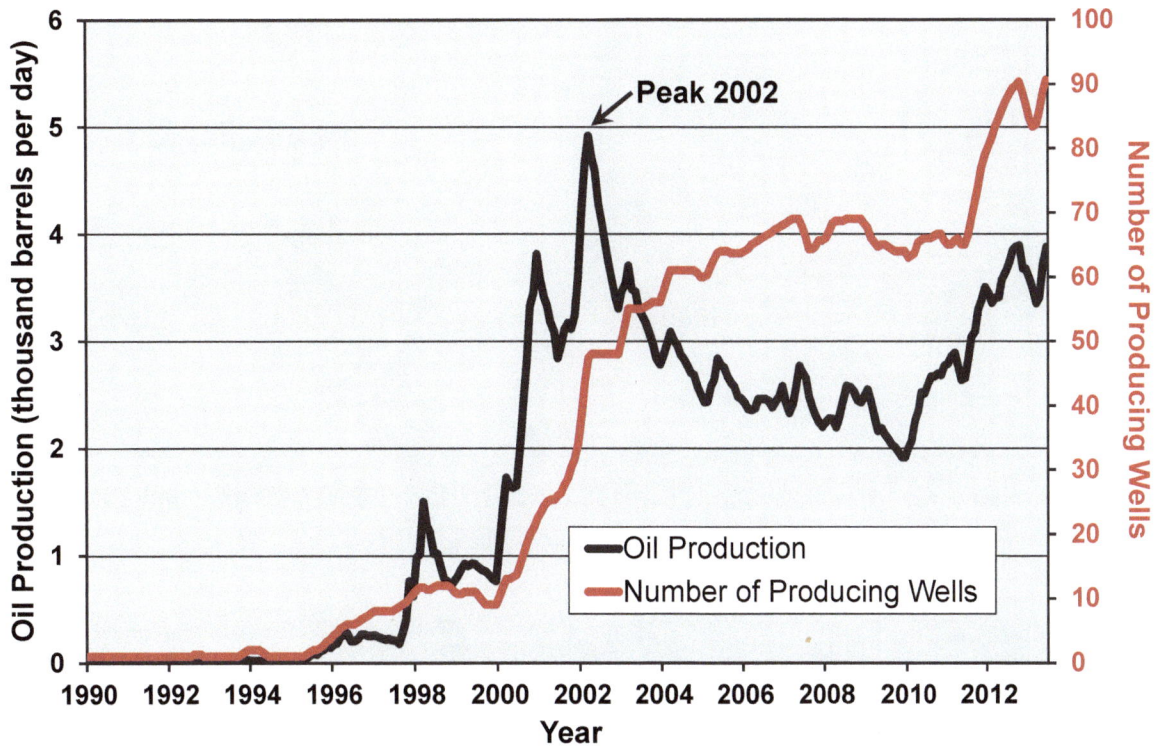

Figure 23. Oil production and producing well count in the North Shafter and Rose Fields, 1990 through June 2013.[45]

Oil production peaked in 2002 but is now rising owing to the increasing number of producing wells.

[45] Data from DI Desktop (Drillinginfo), current through June 2013.

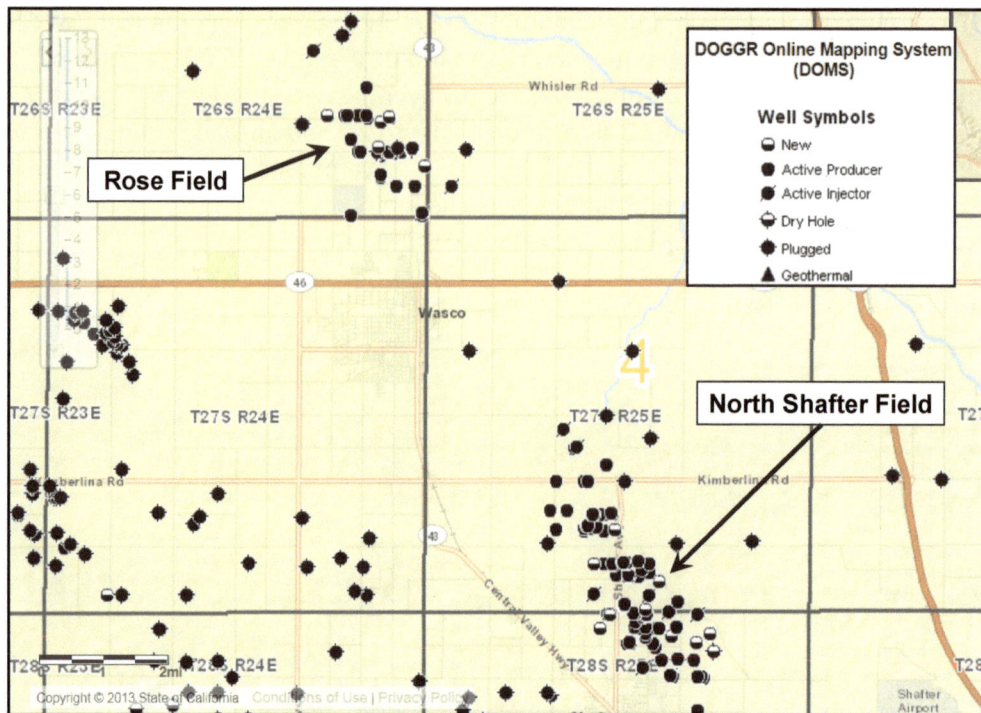

Figure 24. Distribution of wells by status in the Rose and North Shafter Fields as of mid-2013.

Image: California Division of Oil, Gas & Geothermal Resources (DOGGR).[46]

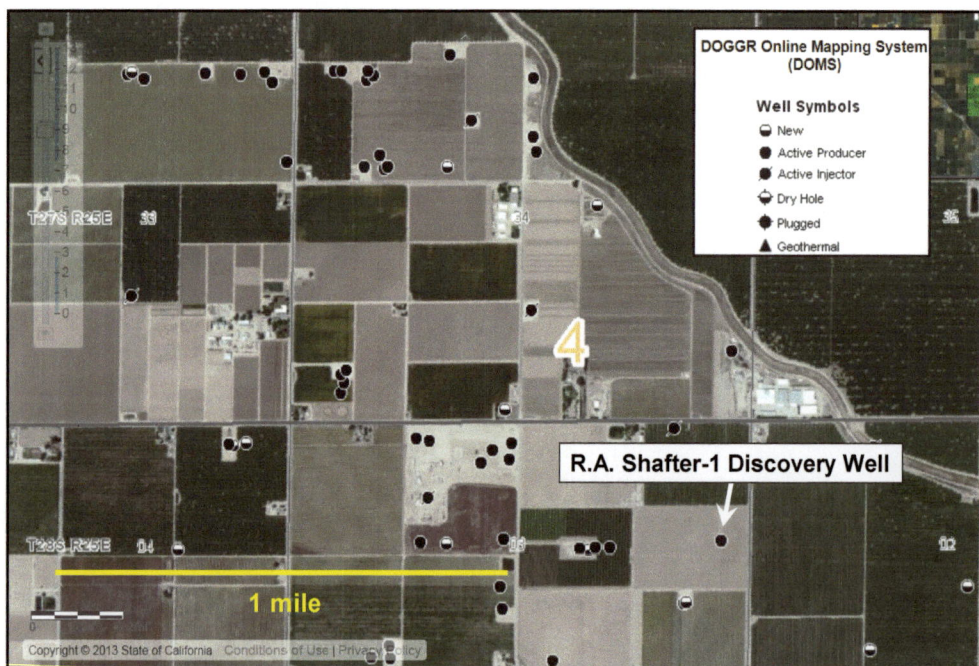

Figure 25. View of central part of North Shafter Field illustrating well footprint as of mid-2013.

The 1983 discovery well is indicated. Well density is over 16 wells per square mile in some cases. Image: California Division of Oil, Gas & Geothermal Resources (DOGGR).[47]

---

[46] California Department of Conservation, Division of Oil, Gas and Geothermal Resources, online mapping system, October 2013, http://maps.conservation.ca.gov/doms/doms-app.html.

A look at well quality in the Rose and North Shafter fields reveals that the best wells (as defined by IP) were drilled early in the development of these fields (Figure 26). This is not surprising as companies usually drill their best prospects first before moving on to less productive prospects. The average IP of directional and vertical wells drilled since 2005 is 157 and 94 barrels per day, respectively. This is far below the average thresholds assumed in the EIA/INTEK report for typical shale wells (500 and 200 barrels per day, respectively), which are indicated in Figure 26.

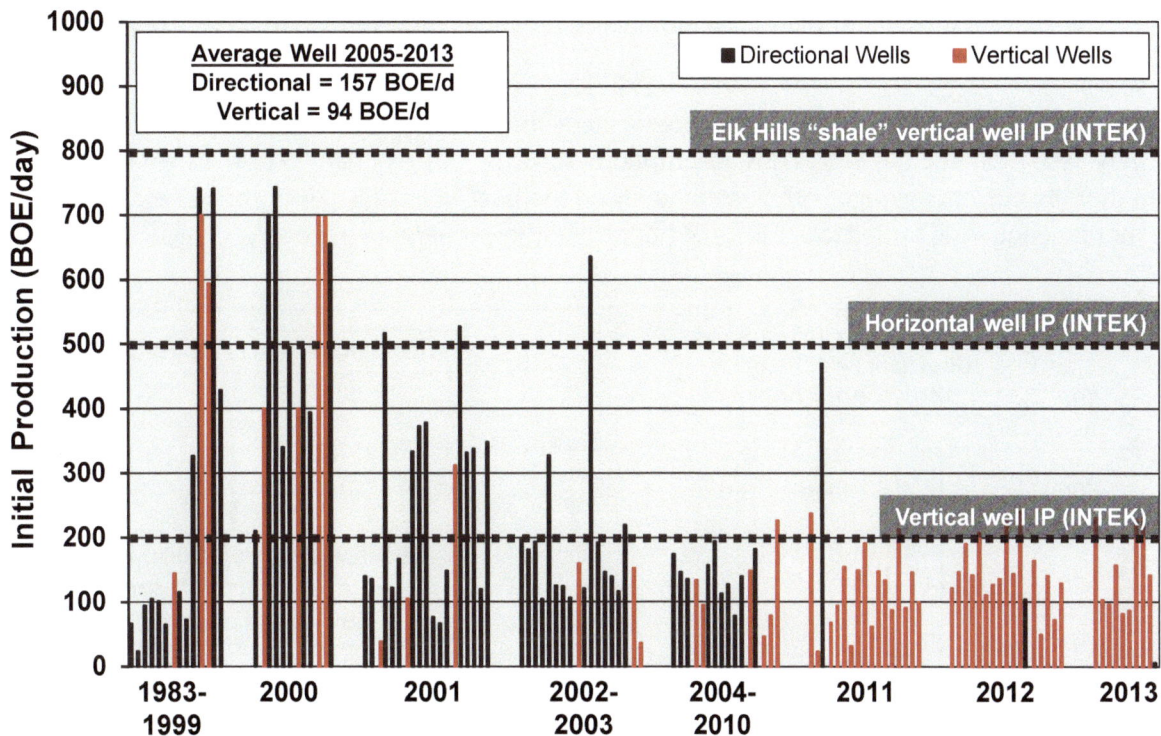

Figure 26. Initial productivity of all Monterey shale wells drilled in the Rose and North Shafter Fields, grouped by year of first production, 1980 through June 2013.[48]

Data include only wells that had oil production of greater than zero barrels of oil equivalent per day. The IPs for "typical" wells as claimed in the EIA/INTEK report (see Figure 19) are indicated.[49] The average IP of directional and vertical wells drilled since 2005 is 157 and 94 barrels per day, respectively. The best wells were discovered early in the development of these fields and recent wells fall far short of the "typical" shale well assumed in the EIA/INTEK report.

[47] California Department of Conservation, Division of Oil, Gas and Geothermal Resources, online mapping system, October 2013, http://maps.conservation.ca.gov/doms/doms-app.html
[48] Data from DI Desktop (Drillinginfo), current through June 2013.
[49] INTEK, Inc., *Review of Emerging Resources: U.S. Shale Gas and Shale Oil Plays*, December 2010, in U.S. Energy Information Administration, *Review of Emerging Resources: U.S. Shale Gas and Shale Oil Plays*, July 2011, http://www.eia.gov/analysis/studies/usshalegas/.

Perhaps the most important metric of the potential of the Monterey shale is how much oil a well will produce over its lifetime—its ultimate production (this is indicated by cumulative production but the ultimate production is not definitively known until the well is shut down). The EIA/INTEK report assumes that each of 28,032 wells drilled at a density of 16 wells per square mile over 1,752 square miles will produce 550,000 barrels of oil on average over its lifespan. Figure 27 illustrates the cumulative production of oil from all wells in the Rose and North Shafter fields. Only two wells out of 123 have exceeded the 550,000 barrel average assumed by the EIA/INTEK report. The average cumulative recovery of oil from wells more than 10 years old for directional and vertical wells is 197,000 and 180,000 barrels, respectively.

Optimistically, currently producing wells in these fields may average 250,000 barrels for ultimate recovery, less than half the EIA/INTEK assumption. However, this estimate is truly a best-case scenario given that the current averages are based mainly on the best wells drilled early on in field development, and more recent wells exhibit lower IPs and hence likely lower ultimate recovery.

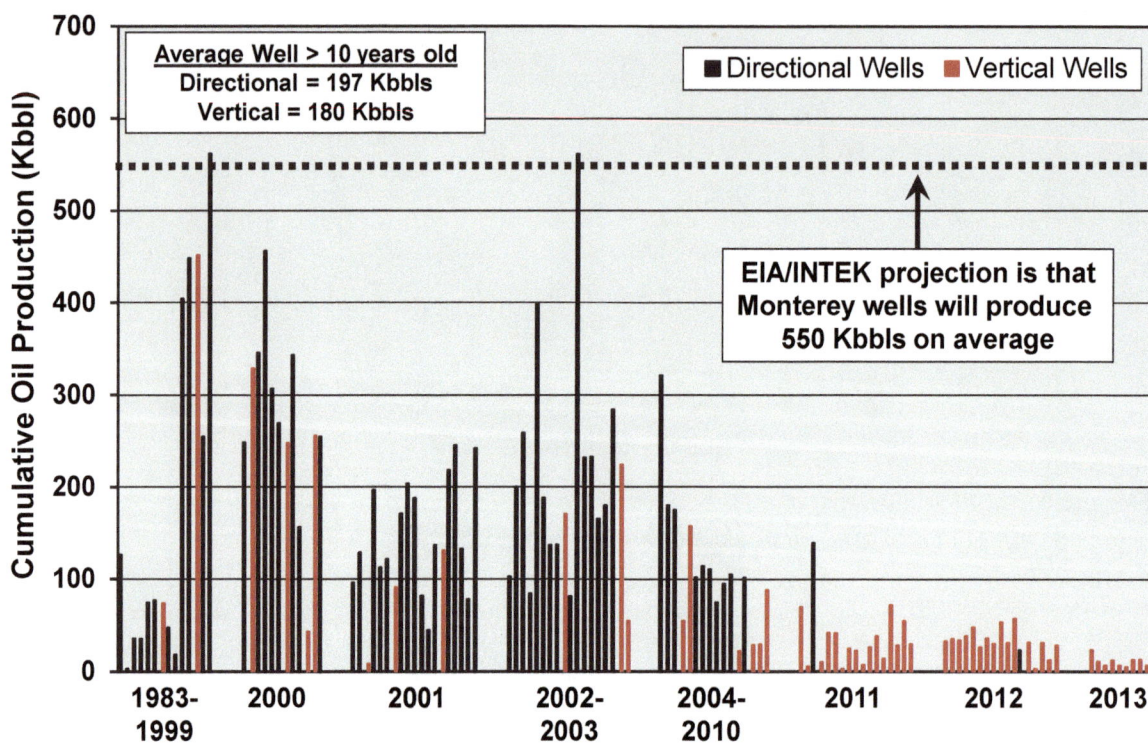

Figure 27. Cumulative oil recovery of all Monterey shale wells drilled in the Rose and North Shafter Fields, grouped by year of first production, 1980 through June 2013.[50]

Data include only wells that had oil production of greater than zero barrels of oil equivalent per day The average per-well cumulative production assumed in the EIA/INTEK report is indicated.[51] The average cumulative oil recovery of wells more than 10 years old is 197,000 barrels for directional wells and 180,000 barrels for vertical wells. It appears unlikely that wells will recover even half as much oil as the 550,000 barrel assumption of the EIA/INTEK report.

[50] Data from DI Desktop (Drillinginfo), current through June 2013.
[51] INTEK, Inc., *Review of Emerging Resources: U.S. Shale Gas and Shale Oil Plays*, December 2010, in U.S. Energy Information Administration, *Review of Emerging Resources: U.S. Shale Gas and Shale Oil Plays*, July 2011, http://www.eia.gov/analysis/studies/usshalegas/.

4.4.3 Production in the San Joaquin Basin

The initial productivity of all shale wells producing from the Monterey Formation in the San Joaquin Basin is illustrated in Figure 28 (this excludes production from the conventional Stevens Sand member of the Monterey). Only four of 866 wells drilled in the last 33 years have exceeded the IP of the typical "Elk Hills Vertical Shale Well" (800 barrels of oil equivalent per day) assumed in the EIA/INTEK report, and all of these were drilled prior to 1995. The most recent data reveal that wells have been unable to produce anything close to the optimistic assumptions of the EIA/INTEK report. The average IPs of directional and vertical wells drilled since 2000 are 169 and 116 barrels of oil equivalent per day, respectively, compared to the EIA/INTEK respective estimates of 500 and 200 barrels of oil equivalent per day. It is also noteworthy that drilling has reverted to mainly vertical wells in the past five years.

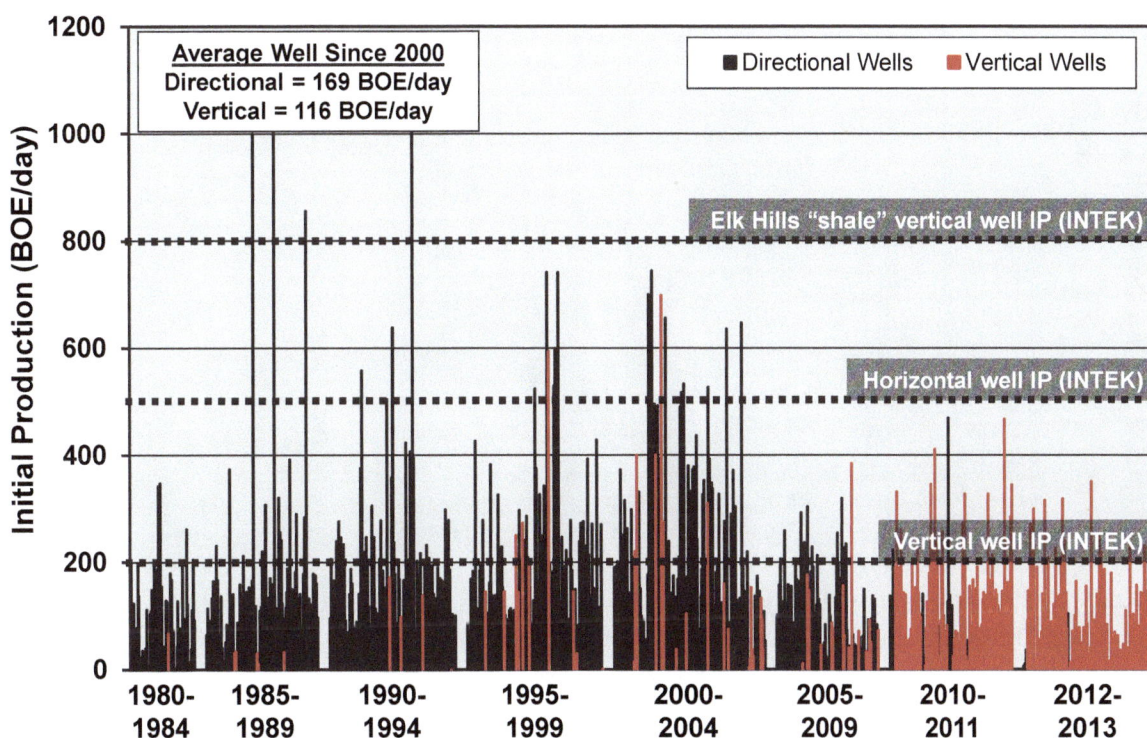

Figure 28. Initial productivity of all Monterey shale wells drilled in the San Joaquin Basin, grouped by year of first production, 1980 through June 2013.[52]

Data include only wells that had oil production of greater than zero barrels of oil equivalent (BOE). The IPs for "typical" wells as claimed in the EIA/INTEK report (see Figure 19) are indicated.[53] Recent wells fall far short of the "typical" shale wells assumed in the EIA/INTEK report; the average IPs of directional and vertical wells drilled since 2000 are 169 and 116 barrels of oil equivalent per day.

---

[52] Data from DI Desktop (Drillinginfo), current through June 2013.
[53] INTEK, Inc., *Review of Emerging Resources: U.S. Shale Gas and Shale Oil Plays*, December 2010, in U.S. Energy Information Administration, *Review of Emerging Resources: U.S. Shale Gas and Shale Oil Plays*, July 2011, http://www.eia.gov/analysis/studies/usshalegas/.

Figure 29 illustrates the cumulative oil production of all shale wells drilled in the Monterey Formation within the San Joaquin Basin since 1980. Only three wells out of 876 have exceeded the average per well ultimate recovery of 550,000 barrels assumed in the EIA/INTEK report. Average cumulative recovery from wells that have been producing for more than 10 years is 97,000 barrels for directional wells and 127,000 for vertical wells—less than one quarter of the average per well ultimate recovery assumed in the EIA/INTEK.

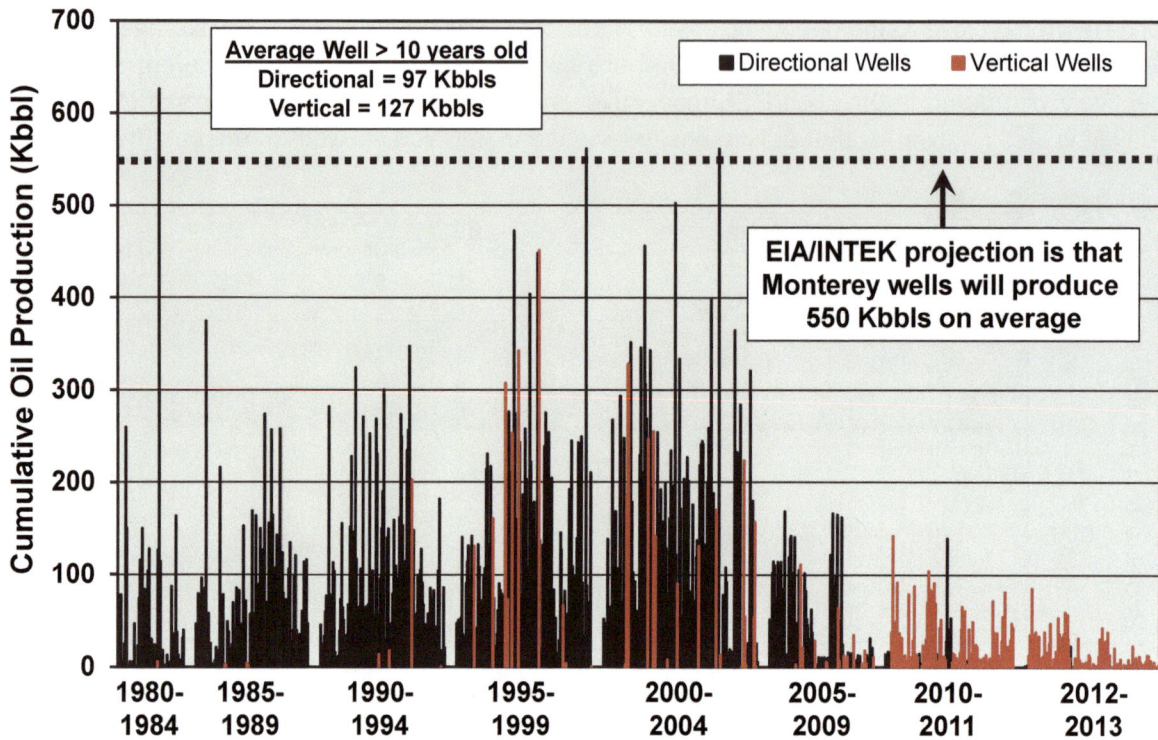

Figure 29. Cumulative oil recovery of all Monterey shale wells drilled in the San Joaquin Basin, grouped by year of first production, 1980 through June 2013.[54]

Data include only wells that had oil production of greater than zero. The average per-well cumulative production assumed in the EIA/INTEK report is indicated. [55] The average cumulative oil recovery of wells more than 10 years old is 97,000 barrels for directional wells and 127,000 barrels for vertical wells. It appears unlikely that wells on average will recover even a third as much oil as the assumption of the EIA/INTEK report.

[54] Data from DI Desktop (Drillinginfo), current through June 2013.
[55] INTEK, Inc., *Review of Emerging Resources: U.S. Shale Gas and Shale Oil Plays*, December 2010, in U.S. Energy Information Administration, *Review of Emerging Resources: U.S. Shale Gas and Shale Oil Plays*, July 2011, http://www.eia.gov/analysis/studies/usshalegas/.

4.4.4 Production in the Santa Maria Basin

The initial productivity of all shale wells producing from the Monterey Formation in the onshore Santa Maria Basin is illustrated in Figure 30. Only one of 113 wells drilled in the last 30 years has exceeded the IP of the typical "Elk Hills Vertical Shale Well" (800 barrels of oil equivalent per day) assumed in the EIA/INTEK report. The most recent data reveal that wells have been unable to produce anything close to the optimistic assumptions of the EIA/INTEK report. Relatively little drilling has occurred in the last three years and the results from that drilling are disappointing, compared to earlier results. The average IPs of directional and vertical wells drilled since 2000 are 116 and 28 barrels of oil equivalent per day, respectively, compared to the EIA/INTEK respective estimates of 500 and 200 barrels of oil equivalent per day.

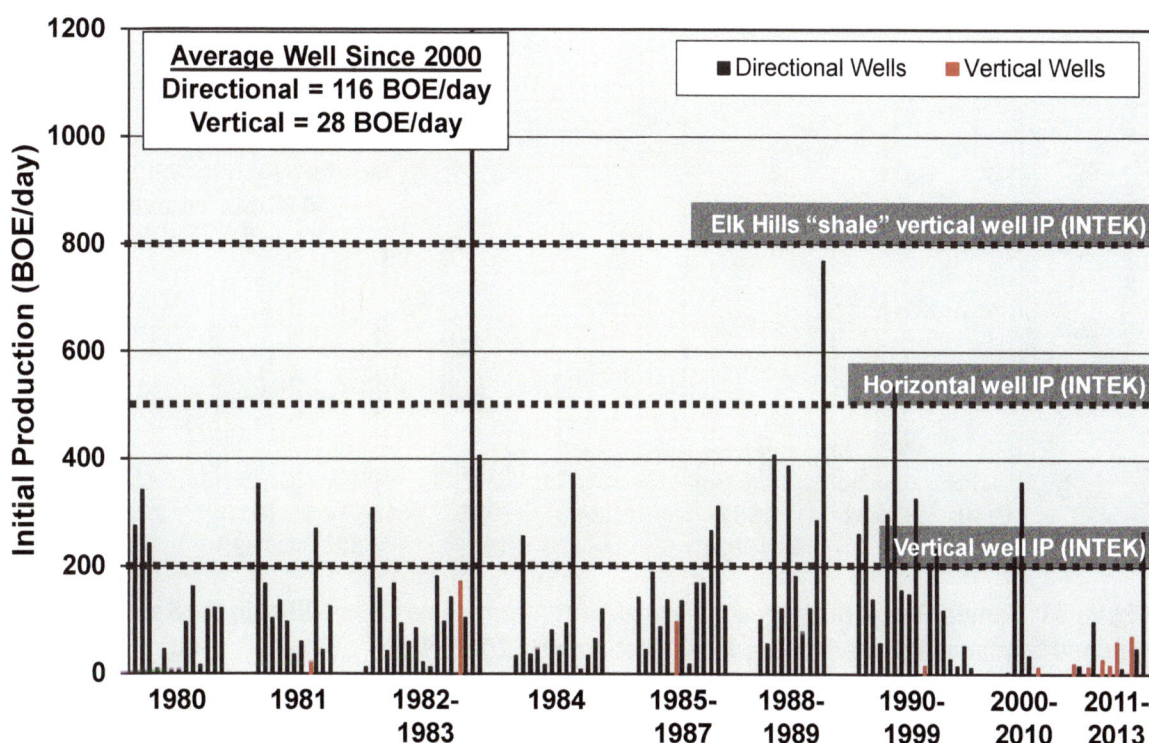

Figure 30. Initial productivity of all onshore Monterey shale wells drilled in the onshore Santa Maria Basin, grouped by year of first production, 1980 through June 2013.[56]

Data include only wells that had oil production of greater than zero barrels of oil equivalent (BOE). The IPs for "typical" wells as claimed in the EIA/INTEK report (see Figure 19) are indicated.[57] Recent wells fall far short of the "typical" shale wells assumed in the EIA/INTEK report; the average IPs of directional and vertical wells drilled since 2000 are 116 and 28 barrels of oil equivalent per day, respectively.

---

[56] Data from DI Desktop (Drillinginfo), current through June 2013.
[57] INTEK, Inc., *Review of Emerging Resources: U.S. Shale Gas and Shale Oil Plays*, December 2010, in U.S. Energy Information Administration, *Review of Emerging Resources: U.S. Shale Gas and Shale Oil Plays*, July 2011, http://www.eia.gov/analysis/studies/usshalegas/.

Figure 31 illustrates the cumulative oil production of all shale wells drilled in the Monterey Formation within the onshore Santa Maria Basin since 1980. Out of 113 wells, only four—all of which were drilled before 1990—have exceeded the average per well ultimate recovery of 550,000 barrels assumed in the EIA/INTEK report. Average cumulative recovery from wells that have been producing for more than 10 years is 141,000 barrels for directional wells and 67,000 for vertical wells—less than one third of the average per well ultimate recovery assumed in the EIA/INTEK report.

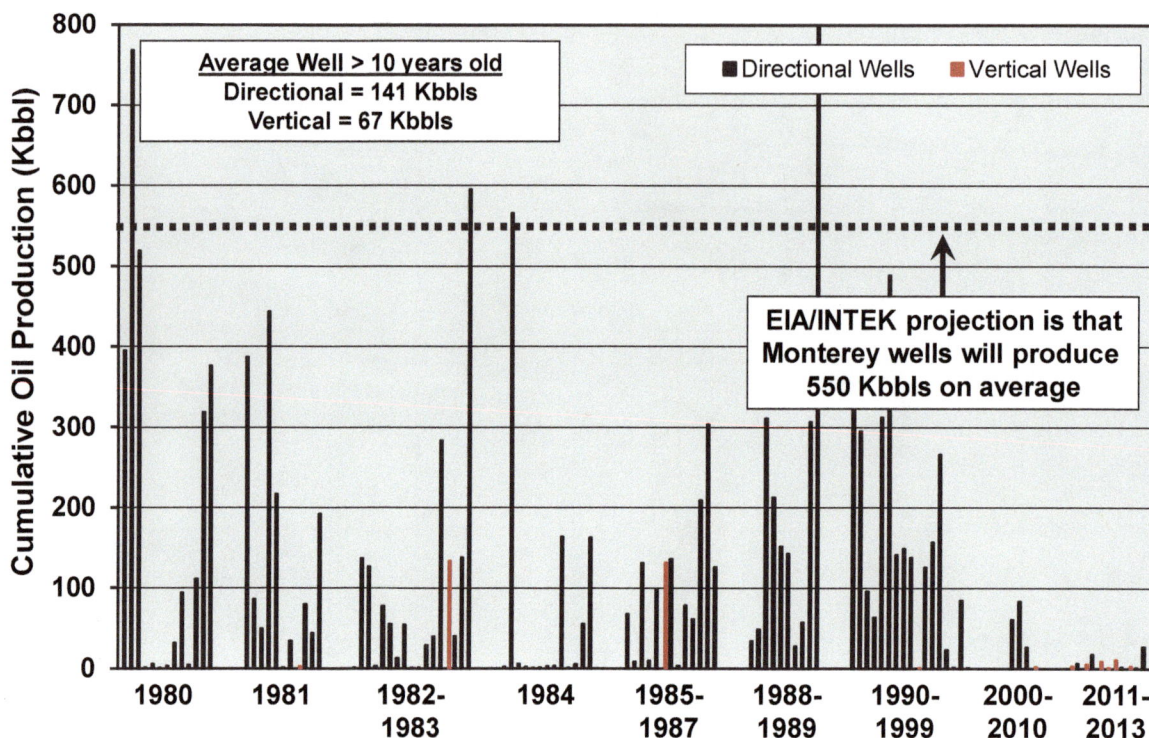

Figure 31. Cumulative oil recovery of all onshore Monterey shale wells drilled in the Santa Maria Basin, grouped by year of first production, 1980 through June 2013.[58]

Data include only wells that had oil production of greater than zero. The average per-well cumulative production assumed in the EIA/INTEK report is indicated.[59] The average cumulative oil recovery of wells more than 10 years old is 141,000 barrels for directional wells and 67,000 barrels for vertical wells. It appears unlikely that wells on average will recover even a third as much oil as the assumption of the EIA/INTEK report.

[58] Data from DI Desktop (Drillinginfo), current through June 2013.
[59] INTEK, Inc., *Review of Emerging Resources: U.S. Shale Gas and Shale Oil Plays*, December 2010, in U.S. Energy Information Administration, *Review of Emerging Resources: U.S. Shale Gas and Shale Oil Plays*, July 2011, http://www.eia.gov/analysis/studies/usshalegas/.

5 REVISITING PRODUCTION FORECASTS OF THE EIA/INTEK REPORT

The preceding review of the geological characteristics of the Monterey shale, together with the analysis of the data on oil production from the Monterey Formation to date, strongly suggest that the EIA/INTEK report's estimate of 15.4 billion barrels of technically recoverable oil is wildly overoptimistic. The analysis reveals the following:

Geology

- The Monterey is not comparable to other tight oil plays. It is structurally and stratigraphically complex and thus is highly variable compared to plays such as the Bakken and Eagle Ford.

- Even within the Monterey's limited areal extent, areas of uplifted mature source rocks with non-migrated oil are likely to be much smaller than assumed by the EIA/INTEK report.

- Production from the Monterey has to date been largely from areally restricted structural and stratigraphic traps charged with migrated oil.

Thus the EIA/INTEK report's basic assumptions about the area potentially available for tight oil production are likely highly optimistic (the report assumes that 28,032 wells can be drilled over a 1,752 square mile area, for a density of 16 wells per square mile). The notion of widespread regions that can be drilled at densities of 16 wells per square mile, with oil production per well at multiples of current well productivity and cumulative production, is wishful thinking that grossly overestimates true oil recovery potential.

Productivity

- Initial productivity per well from existing Monterey shale wells is on average only a half to a quarter of the assumptions in the EIA/INTEK report. Of 979 shale wells drilled in the Monterey, only five had initial productivity above the average "Elk Hills Area Shale Vertical Well" assumed the EIA/INTEK report

- Cumulative recovery of oil per well from existing Monterey wells is likely to average a third or less of that assumed by the EIA/INTEK report. Of 979 shale wells drilled in the Monterey, only seven have had cumulative recoveries above the average assumed the EIA/INTEK report.

- If there were breakthroughs that allowed productivities comparable to the "typical" shale well IPs assumed in the EIA/INTEK report, they should have shown up in the most recent data analyzed for this report. Companies are certainly using the most current technology to stimulate production from new wells, yet growth in the quality of wells (as defined by IP) is stagnant.

- Going forward, it is likely that most development of the Monterey will occur in the San Joaquin Basin of Kern County. There have been relatively few wells drilled recently in the

Santa Maria Basin of Santa Barbara County targeting the Monterey, and environmental concerns of residents there are likely to limit future development.

The San Joaquin and Santa Maria Basins, where the predominant shale potential lies, have been producing from the Monterey for many decades. Certainly advanced 3D seismic analysis will identify other opportunities, but these are likely to be marginal incremental additions, not a new energy bonanza. At best they may help to somewhat offset the inexorable decline in overall California oil production.

Comments such as this from the Deputy Director of the California Conservation Department, which regulates the industry, add confirmation to this analysis:[60]

> *None of the companies that have tried it so far have had significant success, and it doesn't appear to be widespread. It may take an advancement in technology or methodology to unlock the oil production potential of the formation.*

Similarly this comment from Wall Street:[61]

> *"The Monterey shale was supposed to be the greatest thing since sliced bread, but so far has not lived up to the hype," Fadel Gheit, an oil and gas analyst at Oppenheimer & Co. in New York, said in a telephone interview. "It's not conclusive that the emperor has no clothes. So far, it has not shown any big sign that this is going to be another Bakken or Eagle Ford."*

The Monterey shale is highly unlikely to provide an energy bonanza for California on the scale envisioned by the EIA/INTEK report. The EIA/INTEK estimate of 15.4 billion barrels of technically recoverable oil is highly speculative (as is the revised, but largely ignored, 13.7 billion barrel estimate). It is best viewed as an optimistic back-of-the-envelope estimate, rather than something to base energy policy on.

[60] Bloomberg News, "California's Fracking Bonanza may fall short of Promise", April 9, 2013, http://www.bloomberg.com/news/2013-04-10/california-s-fracking-bonanza-may-fall-short-of-promise.html
[61] Chris Reed, "Wall Street doubts CA shale hype – but not Occidental," *CalWatchdog*, April 11, 2013, http://calwatchdog.com/2013/04/11/wall-street-doubts-ca-shale-hype-but-not-occidental/.

6 REVISITING ECONOMIC FORECASTS OF THE USC STUDY

The University of Southern California (USC) published a study in March 2013[62] which stated that "accelerated shale-oil development" in the Monterey Formation would, by 2020:

- Increase per capita GDP in California by $10,300 per year (14.3%).
- Increase statewide employment by 2,815,800 people (10%).
- Increase state and local government tax revenue by $24.6 billion per year (10%).

These are impressive claims that bear closer scrutiny. First and most important, the forecasts of Monterey tight oil production growth upon which these projections were based must be considered (Table 1). The forecasts found that Monterey tight oil production would increase California oil production by 0.18-1.1 million barrels per day in 2020, and by 0.15-3.3 million barrels per day in 2030. In the highest case, this amounts to more than tripling California oil production by 2020 and increasing it seven-fold by 2030 (California reference case oil production is projected by the EIA to be 0.47 mbd in both 2020 and 2030.)[63]

Scenario	Increase in California Oil Production from Monterey Tight Oil (mbd)			
	2015	2020	2025	2030
Decline Curve Analysis (1)	0.16	1.1	1.6	1.6
Decline Curve Analysis (2)	0.15	1.1	2.4	3.3
Modified Decline Curve Analysis (1)	0.15	0.69	0.67	0.42
Modified Decline Curve Analysis (2)	0.077	0.58	1.2	1.7
Incremental Advanced-Technology Oil Production	0.13	0.18	0.17	0.15
Average of all scenarios	0.1334	0.73	1.208	1.434
Average of highest and lowest scenarios	0.14	0.64	1.285	1.725

Table 1. Scenarios of increases in oil production from the Monterey shale in the USC study, 2015-2030.

The "Decline Curve Analysis 1" and "Decline Curve Analysis 2" scenarios utilize the production curves from wells "3-1" and "RA-1" in the North Shafter Field, respectively. The "Modified Decline Curve Analysis" scenarios apply a "degradation factor" to these two scenarios.[64] The "Incremental Advanced-Technology Oil Production" scenario is the minimum assumed increase in California oil production from the Monterey over the EIA reference projections.[65] Depending on assumptions, Monterey production is forecast to grow by between 0.18 and 1.1 million barrels per day by 2020.[66]

Although the USC study does not specifically indicate the actual oil production figures used to calculate the estimated impact of Monterey production on California, it does state: *"It therefore makes the most*

---

[62] University of Southern California Global Energy Network, *The Monterey Shale and California's Economic Future*, (March 2013), http://gen.usc.edu/assets/001/84955.pdf.

[63] As noted in Table K1, University of Southern California Global Energy Network, *The Monterey Shale and California's Economic Future*, (March 2013), http://gen.usc.edu/assets/001/84955.pdf.

[64] See Table K3 in University of Southern California Global Energy Network, *The Monterey Shale and California's Economic Future*, (March 2013), http://gen.usc.edu/assets/001/84955.pdf.

[65] See Table K1 in University of Southern California Global Energy Network, *The Monterey Shale and California's Economic Future*, (March 2013), http://gen.usc.edu/assets/001/84955.pdf.

[66] University of Southern California Global Energy Network, *The Monterey Shale and California's Economic Future*, (March 2013), http://gen.usc.edu/assets/001/84955.pdf; see Appendix K, page 68. Although the original table title refers to resources "in California," the text clearly introduces the table as a forecast for the Monterey.

sense to emphasize a set of median (half-way) scenario results."[67] Presumably this means that the average of all Monterey oil production scenarios as listed in Table 1 is the basis for the USC economic analysis. This would require an astonishing 155 percent overall growth in California oil production over EIA projections by 2020 and a 305 percent growth by 2030.

Figure 32 graphically illustrates the USC Monterey oil production scenarios for California in Table 1. The analysis of the Monterey tight oil play summarized in Section 1 found no basis for forecasts of such significant future production growth; thus to say that these projections are optimistic would be an understatement. Furthermore, the USC study suggests that these projected increases in oil production are possible by 2020 with just 1,380 new wells, and by 2030 with just 4,112 wells (Figure 33). The average USC study scenario in Table 1 would require each new well to produce 529 barrels per day in 2020, which is more than ten times higher than the current average per-day production of wells in the highest producing parts of the Monterey shale.

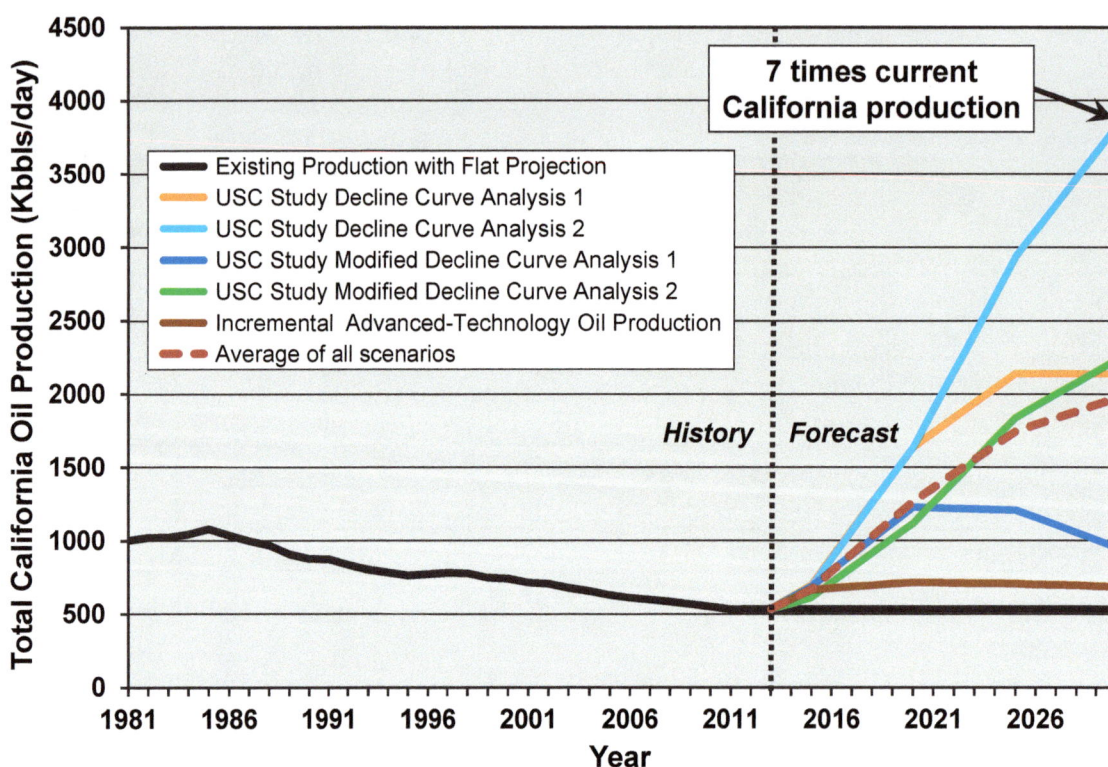

Figure 32. USC study assumptions of growth in Monterey tight oil production, 2013-2030.

Figure is based on the USC study scenarios given in Table 1. The high case would see production grow seven-fold through 2030, and the average of all scenarios would see California production more than double by 2020.

[67] See page 3 in University of Southern California Global Energy Network, *The Monterey Shale and California's Economic Future*, (March 2013), http://gen.usc.edu/assets/001/84955.pdf.

Table K2. New well drilling plan in Monterey Shale Formation										
Year	2013	2014	2015	2016	2017	2018	2019	2020	2021	2022
Wells to be drilled	32	80	116	152	168	268	284	280	272	276

Year	2023	2024	2025	2026	2027	2028	2029	2030	2031	2032	2033	2034	2035
Wells to be drilled	288	288	288	248	268	268	268	268	268	268	268	268	268

Figure 33. USC study estimates of the number of wells required for Monterey tight oil production growth indicated in Table 1 and Figure 32.[68]

Table K2 in the USC study, displayed as in the original. The high case would see production grow seven-fold through 2030, and the average of all scenarios would see California production more than double by 2020.

The USC economic study utilized two wells in the North Shafter Field to estimate the future production projections for the Monterey shale. Well RA-1 is the discovery well for the North Shafter Field, and has been producing since 1983. Well 3-1 was drilled in 1995. The production profiles for these wells are illustrated in Figure 34.

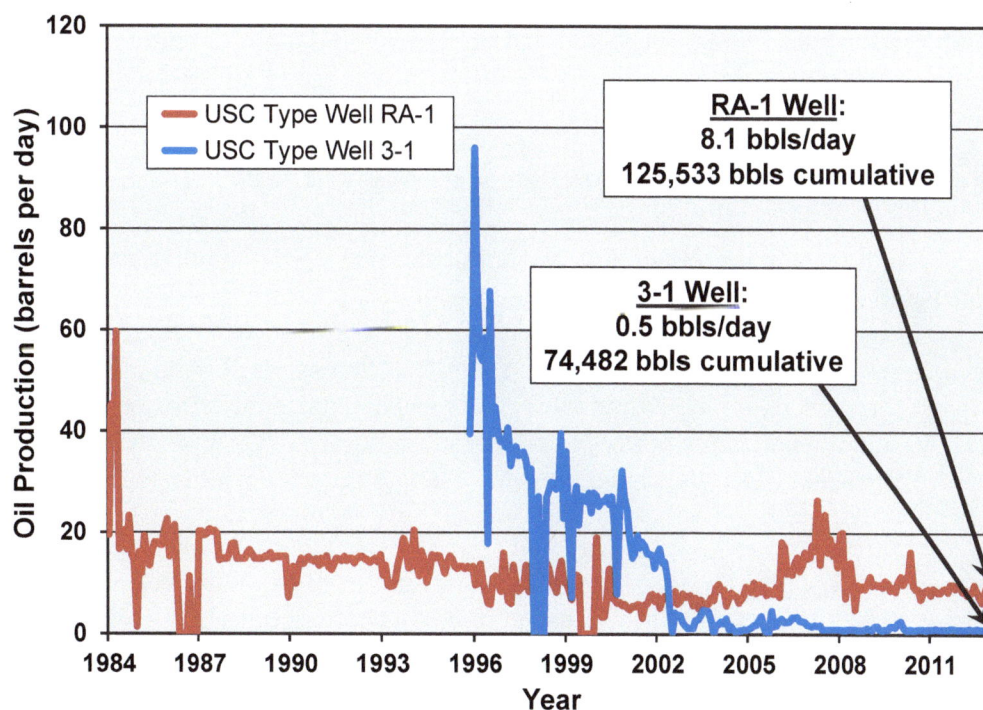

Figure 34. Reference well production used by the USC study to estimate Monterey tight oil production in various scenarios, 1984 through February 2013.[69]

Figure is based on the scenarios given in Table 1 (see also Figure 32). These wells are both from the North Shafter Field which is producing from the McLure Shale member of the Monterey.

---

[68] See Table K2 in University of Southern California Global Energy Network, *The Monterey Shale and California's Economic Future*, (March 2013), http://gen.usc.edu/assets/001/84955.pdf.
[69] Data from DI Desktop (Drillinginfo), August 2013.

The USC economic study has radically underestimated the number of wells required if average well production corresponds to the two wells from the North Shafter Field it used to make its estimates. Table 2 is a decline curve analysis of the number of wells that would actually be required to meet the 2030 production forecasts in Table 1 and Figure 32 using the two reference North Shafter wells. The actual number of wells required would be between 49,119 and 232,562, compared to the 4,112 assumed in the USC study.

USC Scenario	USC Reference Well	USC 2030 Monterey Production (million barrels per day)	USC Number of Wells	Actual Number of Wells Required to Meet USC Forecast with USC Reference Well	USC Oil Recovery through 2030 (billion barrels)
Decline Curve Analysis 1	3-1	1.6	4,112	117,269	6.89
Decline Curve Analysis 2	RA-1	3.3	4,112	232,562	10.20
Modified Decline Curve Analysis 1	3-1	0.42	4,112	49,119	3.14
Modified Decline Curve Analysis 2	RA-1	1.7	4,112	119,532	5.21

Table 2. USC study assumptions of the wells required to meet 2030 production estimates compared to the actual number of wells required.

Actual number of wells required is based on an analysis of decline curves for the reference wells used in the USC study (see also Figure 34).[70] The highest oil production scenario would see the recovery of some 10.2 billion barrels by 2030, or 66 percent of the purported 15.4 billion barrels of technically recoverable Monterey tight oil resources in the EIA/INTEK report.

Considering that some 50,000 wells are required for current California oil production, estimates of more than 200,000 additional wells to grow California production seven-fold (USC's highest case) are not unreasonable. One might argue that the reference wells chosen by the USC were not representative, and production estimates should have been higher. Even so, tens of thousands of wells would have to be drilled to meet the USC forecasts, far more than the 4,112 wells it considered were necessary by 2030 in its report. Given that the estimates of available locations by the EIA/INTEK report are at most 28,032, and likely to be much less given the above analysis, the oil production estimates in the USC study lack credibility.

The USC study based some economic assumptions on the history of recent economic growth in North Dakota related to the Bakken oil boom. Curiously, however, it also utilized South Dakota and Wyoming— states which have seen some recent growth in oil production, but at extremely small scales, and not related to tight oil as in the Monterey:

> We conducted analysis for California as well as three oil boom states: North Dakota, Wyoming and South Dakota. These three states have recently experienced the effects of a significant oil drilling boom.[71]

---

[70] University of Southern California Global Energy Network, *The Monterey Shale and California's Economic Future*, (March 2013), http://gen.usc.edu/assets/001/84955.pdf.

A recent study funded by the American Petroleum Institute found that unconventional oil and gas development in the entire United States supported 2.1 million jobs in 2012.[72] The USC economic report suggests that the Monterey in California alone could support 2.8 million jobs by 2020.

In summary, the USC study of the potential economic impact of developing the Monterey shale lacks credibility. The oil production forecasts on which it is based are extremely optimistic and are likely impossible to achieve, given the realities of the geology, production potential, and the number of drilling locations in the Monterey shale as described in this report.

---

[71] University of Southern California Global Energy Network, *The Monterey Shale and California's Economic Future*, (March 2013), http://gen.usc.edu/assets/001/84955.pdf; see page 28.
[72] American Petroleum Institute, "Landmark study shows manufacturing renaissance from U.S. shale," Press Release, September 4, 2013, http://www.api.org/news-and-media/news/newsitems/2013/sept-2013/landmark-study-shows-manufacturing-renaissance-from-us-shale.

7 SUMMARY AND CONCLUSIONS

California oil production has been in decline since it peaked in 1985 at 1.1 million barrels per day. Current production is just over half a million barrels per day, a decline of 50 percent. The Monterey shale has been a very important contributor to California oil production both as a source rock—which charged many conventional oil fields—and to a lesser extent as a reservoir rock.

New technology of high-volume, multi-stage hydraulic fracturing of horizontal wells, which has resulted in rapidly growing production from tight oil plays elsewhere in the U.S., has sparked hope of a resurgence of California oil production from the Monterey shale. It has been suggested in the 2011 EIA/INTEK report that the Monterey shale is the largest tight oil play in the U.S., with technically recoverable resources of 15.4 billion barrels, or 64 percent of Lower-48 U.S. tight oil potential. This widely-cited report assumed that broad regions, totalling 1,752 square miles, will prove productive and can be drilled at a density of 16 wells per square mile, with each well recovering 550,000 barrels of oil.

The detailed analysis of geological and well production data in this report suggests this assessment is simplistic and likely highly overstated:

- Existing fields within the Monterey are areally restricted and are primarily controlled by structural and stratigraphic trapping mechanisms, thus the assumption of broad regions of prospectivity is highly questionable.

- The Rose and North Shafter fields, which may be the closest approximation to a "prototype" for the Monterey tight oil play, are not laterally extensive and were discovered by careful seismic mapping, as were most of the other producing Monterey fields.

- There are few wells even close to the 800 barrels of oil equivalent production cited for the initial productivity of the "typical Elk Hills vertical shale well" used as a basis for assumptions of production potential in the EIA/INTEK report. Furthermore, the Elk Hills is primarily producing from a sand member of the Monterey, not a shale, and hence is not an analogue for what to expect from a Monterey shale reservoir.

- An analysis of every well producing from Monterey shale reservoirs reveals that average initial productivity is less than half of the typical horizontal and vertical shale wells assumed in the EIA/INTEK report, and less than a quarter of the "typical Elk Hills vertical shale well".

- Notwithstanding that producing Monterey fields are areally restricted (not ubiquitous), average cumulative oil recovery per well is likely to be one-third or less of the average assumed for all wells in the EIA/INTEK report.

- Fracking and acidization have doubtless been tried extensively on Monterey shale wells, yet the data do not show any significant increase in initial well productivity or likely cumulative oil recovery for recent wells.

- The majority of oil produced from the Monterey appears to have migrated, owing to the fractured nature of much of the Monterey. The existence of very extensive areas of uplifted mature source rock with non-migrated oil comparable to plays like the Bakken is highly speculative.

Compounding the unfounded optimism in the EIA/INTEK report is the projection of Monterey-fuelled economic growth in the USC economic study released in March 2013. This study projects between a thirty percent- and a seven-fold increase in California oil production by 2030 as a result of development of the Monterey shale. The study suggests this production growth could be achieved with a mere 4,112 wells, compared with the approximately 50,000 wells from which current California oil production comes. The study further suggests that Monterey development will provide $24.6 billion per year in tax revenue, and provide 2.8 million jobs by 2020, which is more than are currently employed in the entire U.S. unconventional oil and gas industry. This study is highly optimistic and lacks credibility in the light of a detailed analysis of the Monterey shale's potential.

Californians would be well advised to avoid thinking of the Monterey shale as a panacea for the State's economic and energy concerns. Although tight oil production has been a temporary game changer in terms of reversing U.S. production declines, the U.S. remains the second largest oil importer in the world. The longer term prognosis for the sustainability of oil production from tight oil plays remains questionable, given high well- and field-production declines that must be overcome with more drilling. The high levels of drilling required to sustain tight oil plays also imply extensive collateral environmental impacts. California has made considerable progress in alternative energy deployment and reducing energy throughputs. This must be aggressively pursued in the future.

GLOSSARY

Burial depth —The current depth at which a reservoir is found, which may be considerably shallower than the maximum burial depth due to subsequent uplift. The maximum burial depth is typically where maximum temperatures and pressures are reached.

Crude oil — As used in this report, conventional crude oil plus other petroleum liquids but not including biofuels or refinery gains.

Cumulative recovery — Total production to date. See also: "Ultimate recovery."

Directional well — A well drilled at an angle to access targets horizontally offset from the drilling pad location.

Formation – A formal name for a rock unit which may be a subdivision of a Group. A Formation may be further subdivided into Members.

Horizontal well — A well typically started vertically which is curved to horizontal at depth to follow a particular rock stratum or reservoir.

Initial Productivity (IP) —The average daily production in the highest month of production from a well, which is normally the first or second month after a well is drilled.

Generation window — An interval in the subsurface where temperatures and pressures are high enough for organic matter (kerogen) to undergo thermogenic breakdown (cracking), generating oil or gas. The oil window is often found in the 60-120 degree Celsius interval (approx. 2-4 km depth), while the corresponding gas window is found in the 100-200+ degree Celsius interval (3-6 km depth).

Hydraulic fracturing ("fracking") — The process of inducing fractures in reservoir rocks through the injection of water and other fluids, chemicals and solids under very high pressure.

Member — A formal name for a rock unit which is a subdivision of a Formation.

Multi-stage hydraulic-fracturing — Each individual hydraulic fracturing treatment is a "stage" localized to a portion of the well. There may be as many as 30 individual hydraulic fracturing stages in some wells.

Oil equivalent — Hydrocarbons, typically gas, converted into the equivalent amount of oil on an energy content basis.

Oil shale — Organic-rich rock that contains kerogen, a precursor of oil. Depending on organic content it can sometimes be burned directly with a calorific value equivalent to a very low grade coal. Can be "cooked" in situ at high temperatures for several years to produce oil or can be retorted in surface operations to produce petroleum liquids.

Play — A prospective area for the production of oil, gas or both. Usually a relatively small contiguous geographic area focused on an individual reservoir.

Reserve — A deposit of oil, gas or coal that can be recovered profitably within existing economic conditions using existing technologies. Has legal implications in terms of company valuations for the Securities and Exchange Commission.

Reservoir rock — See "Source rock."

Resource — See "Technically recoverable resource" and "Undiscovered technically recoverable resource"

Shale gas — Gas contained in shale with very low permeabilities in the micro- to nano-darcy range. Typically produced using horizontal wells with multi-stage hydraulic fracture treatments.

Shale oil — See "tight oil."

Source rock — A rock rich in organic matter which, if heated sufficiently over time, will generate oil or gas. Typical source rocks, usually shales or limestones, contain about 1% organic matter and at least 0.5% total organic carbon (TOC), although a rich source rock might have as much as 10% organic matter.

Subdivision — In rock nomenclature, a unit with specific rock characteristics that allow it to be mapped in surface outcrop and in the subsurface over relatively large geographic areas, one level of which is formally termed a Formation. Formations may be subdivided into Members or aggregated into Groups.

Technically recoverable resource — An oil and/or gas resource that is believed to be recoverable with existing technology with no consideration of the cost of extraction.

Tight oil — Also referred to as shale oil. Oil contained in shale and associated clastic and carbonate rocks with very low permeabilities in the micro- to nano-darcy range. Typically produced using horizontal wells with multi-stage hydraulic fracture treatments.

Type decline curve — The average production declines for all wells in a given area or play from the first month on production. For shale plays in this study the type decline curves considered the average of the first four to five years of production.

Ultimate recovery — Total production as of shut-down of a well. See also: "Cumulative recovery."

Undiscovered technically recoverable resource — Resources inferred to exist using probabilistic methods extrapolated from available exploration data and discovery histories. Usually designated with confidence levels. For example, P90 indicates a 90 percent chance of having a least the stated resource volume whereas a P10 estimate has only a 10 percent chance.

www.ingramcontent.com/pod-product-compliance
Lightning Source LLC
Chambersburg PA
CBHW052054190326

41519CB00002BA/219